Advance Praise for

Building Without Borders

Building Without Borders is the most important recent book on
sustainable building, because it's about serving those in greatest need of
inexpensive, efficient, and culturally and locationally appropriate housing — the
world's poorest billions. From Manila to Mexico City, Calcutta to Caracas, Nairobi
to Rio de Janeiro, vast teeming slums shame us all and dim the human prospect.
Building Without Borders shows practical ways to make the world
better and safer, one dwelling at a time.

—Amory Lovins, CEO, Rocky Mountain Institute and
sustainable home owner/designer

Building Without Borders shows, with an elegant and understated
flourish, that housing the world's burgeoning population is not an intractable
problem — that the poor and displaced can and do house themselves with
affordable solutions, and that the best of these have a sensitivity to the
human ecology that points us all to a better future.

— Albert Bates, President, Global Ecovillage Network

The book we all have waited for. Quite unique in development literature, and
full of thought-provoking case studies, applicable to our own housing problems.
Empowering and encouraging.

— Ianto Evans is owner/director of the North American School of Natural Building
and Cob Cottage Company, co-author of *The Hand-Sculpted House* and
Helping People in Poor Countries

Rarely does one movement address so many needs so elegantly.
In *Building without Borders*, an articulate, caring community of
natural builder/activists recounts everything from the housing and ecological
problems to be addressed, to the nuts and bolts (or mud and straw) of the
building techniques employed, to the complex social issues of how
white Americans can work effectively with indigenous people worldwide,
and ultimately the satisfaction of housing people while teaching them
skills they can use to house others. The world is a richer place
for the work that is chronicled here, and my own
world is richer for having read about it.

— CAROL VENOLIA, eco-architect, *Natural Home Magazine* columnist,
and author of *Healing Environments*

This book is truly inspirational and a great source of information
for those wanting to enter the world of development and low cost sustainable
building. As a natural building development worker, I have been waiting for
a book like this to again feel reconnected with all the activities going
on in this field. It would be fantastic if there were a new volume
that came out every decade to present the contemporary
problems of this forever changing global crisis.

— PAULINA WOJCIECHOWSKA, author, *Building With Earth,*
www.EarthHandsAndHouses.org

Building Without Borders

Building Without Borders

SUSTAINABLE
CONSTRUCTION
FOR THE
GLOBAL VILLAGE

Edited by Joseph F. Kennedy

NEW SOCIETY PUBLISHERS

*For my wife Rose
and daughter Taya*

Cataloguing in Publication Data:

A catalog record for this publication is available from the National Library of Canada.

Cover design by Diane McIntosh. Cover photograph ©Bill Steen, The Canelo Project.

Printed in Canada by Transcontinental Printing.

New Society Publishers acknowledges the support of the Government of Canada through the Book Publishing Industry Development Program (BPIDP) for our publishing activities.

Paperback ISBN: 0-86571-481-9

Inquiries regarding requests to reprint all or part of *Building Without Borders* should be addressed to New Society Publishers at the address below.

To order directly from the publishers, please add $4.50 shipping to the price of the first copy, and $1.00 for each additional copy (plus GST in Canada). Send check or money order to:

New Society Publishers
P.O. Box 189, Gabriola Island, BC V0R 1X0, Canada
1-800-567-6772

New Society Publishers' mission is to publish books that contribute in fundamental ways to building an ecologically sustainable and just society, and to do so with the least possible impact on the environment, in a manner that models this vision. We are committed to doing this not just through education, but through action. We are acting on our commitment to the world's remaining ancient forests by phasing out our paper supply from ancient forests worldwide. This book is one step towards ending global deforestation and climate change. It is printed on acid-free paper that is **100% old growth forest-free** (100% post-consumer recycled), processed chlorine free, and printed with vegetable based, low VOC inks. For further information, or to browse our full list of books and purchase securely, visit our website at: www.newsociety.com

NEW SOCIETY PUBLISHERS   www.newsociety.com

Contents

Preface

The Members of Builders Without Borders
Edited by Owen Geiger

Builders Without Borders (BWB) is a nonprofit organization dedicated to serving the underhoused. We are an international network of ecological builders with the mission to form partnerships with communities and organizations around the world to create affordable housing from local materials, and to work together for a sustainable future. We believe that the solution to homelessness is not merely housing, but a local population trained to provide housing for themselves.

Builders Without Borders believes that within the problem lies the solution. Crisis can afford an opportunity to create something positive and healing. Our view is that we can best work in partnership with local populations to identify needs and propose solutions, and through our involvement leverage important information and material resources. In this vein we seek solutions using local skills and resources and involving all members of society — men and women, young and old, and people of various ethnic origins and different levels of society. BWB also provides workshops, training programs, and educational materials.

Builders Without Borders does not prescribe one solution for all situations, but rather uses a relationship-based process that allows local people and other participants to find unique solutions through collaboration and dialog for the particular problems at hand. Such a process is:

- Empowering. Co-creating building designs and technologies within communities spreads knowledge more widely and effectively.
- Hands-on. Practical skills can be learned by anyone.
- Cross-fertilizing. Creating opportunities for cross-cultural communication

and the demonstration of innovative and traditional building techniques serves to enrich the overall knowledge base.

- Regionally appropriate. The use of locally available knowledge, skills, and materials is emphasized.
- Respectful. Points of cultural difference are respected and points of convergence highlighted.

BWB training emphasizes a sustainable approach to housing development using natural building methods. Natural building makes use of locally available materials such as straw, stone, and earth as opposed to costly imported materials such as milled lumber, steel, and concrete. The advantages of natural building include low costs, easily obtainable materials (which is especially important in areas with limited resources and infrastructure), energy efficiency, low toxicity, safety, and durability. Structures built with natural building materials are typically more readily accepted by local populations and naturally blend in to the vernacular architecture. In addition natural building uses existing work forces of adobe workers, stone masons, plasterers and others and thus offers significant time and cost savings over other methods. This approach is empowering and develops self-sufficiency, with less reliance on outsiders.

Our members and network include professional engineers, architects, builders, and educators on the cutting edge of alternative technologies, many of whom have contributed to this book.

Foreword

Judy Knox

> *"We are all implicated in each others' fates in significant ways and bear tremendous responsibility for that."* — Russell Banks

Each of us who contributes to this book, reads it, and/or hopefully acts from it, is already a community member — bound together by some level of awareness of our interconnectedness with all living beings on this fragile, wondrous, wounded planet. We are not surprised by the knowledge that every living system on Earth is in decline. We accept responsibility that it is the cumulative effect of our anthropocentric, disconnected human choices that have brought us all to this precarious time in the history of our planet.

We care deeply — Oh, that we could heal all the wounds with just our caring! And even though the relative comfort of our lives insulates us from the day-after-day realities of the grinding poverty of millions of our brothers and sisters, and the often silent holocaust of our natural world, we feel the increasing weight and despair that comes from knowing (and caring about) the truth but not yet recognizing the ways we can "response-ably" act on that truth. It is our deep, insistent yearning to close that gap which urges us to undertake the most meaningful journey of our lives — the journey toward a right livelihood that champions those things we most care about.

The journey of a champion is a difficult one, necessitating deep reflection on the ways in which we live our daily lives and recognition of how even our smallest choices might diminish rather than affirm the whole living organism of the Earth and its inhabitants. It means making choices and changes, bit by bit, day by day — changes in how we invest our time, how we interact with others, how we meet our basic human needs, how we perceive ourselves in relation to the whole — until more and more of our lives become a demonstration of the world we wish to create. As we move we become part of a growing community of fellow journeyers, encouraging and cajoling and teaching and enriching each other. We learn that together we can infuse hope into our distressed world — not false hope, but hope that rises out of discovering real and doable ways to meet our basic human needs while restoring and sustaining life.

The ever growing, worldwide sustainable building community has been on the journey for many years now. We have increasingly accepted responsibility for the destructiveness of how we house ourselves in the so-called developed part of the world. Even though we have a long way to go in making the necessary dramatic move — from "egocentric" to "ecocentric" buildings — our community has, in the course of its journey, developed replicable, life-affirming tools that have been generously shared among ourselves and others. Many people around the world have been inspired to build with varying combinations of earth and fibers such as straw bales, cob, adobe, earth bags, bamboo, thatch, and other regionally available, renewable materials. Our "toolkits" also contain important information about how, how much, and where to build — joyfully, as a fully participating family or community.

Matts Myhrman and I have a favorite photo that we use in our public slideshows. It is an aerial view of a huge field of baled straw in the Ukraine. In the foreground a family has just completed digging a foundation for their cement block home, unaware that they could just (as Matts loves to say) "stand the field up on end." We call that photo "The Field of Unrealized Dreams." It represents to us the vast, unfinished work of transferring straw bale construction and many other simple, sustainable building techniques to those around the world who need simple, safe, comfortable, and affordable shelter.

Our common humanity compels us to discover, with great humility, the most appropriate ways to enter and honor unique cultures and traditions while working with these communities to put our most effective tools to work for their benefit. *This* is where the unfulfilled potential of our life-affirming technologies will be realized. *Building Without Borders* is a direct response to that crucial challenge. It is a compendium of the experiences and wisdom of those on the front lines who share, work, and learn with communities in other cultures and countries to do the most good for their inhabitants while doing the least harm to the planet. This book invites any one of us to join in the journey with this thoughtful and dedicated community of way-showers. Please know that we are each essential to this work. Though staying on this path is not the easiest way to live, it gives us access to a life rich with connection, joy, and meaning.

Judy Knox lives and works with her husband and partner Matts Myhrman in the urban center of Tucson, Arizona. Through their business, Out On Bale, they have been working full-time since 1989 to gather, grow, and distribute excellent straw bale construction resources in order to encourage its widespread acceptance and use as a sustainable building method. They publish Build It With Bales, *and founded and published* The Last Straw Journal *until 1998.*

Judy takes great joy in her four grandchildren, and speaks and works for the Earth they will co-inhabit.

Acknowledgments

*B*uilding *Without Borders* is the result of several years of effort by many people. The original research for this book was made possible by a generous grant from The Graham Foundation for Advanced Studies in the Fine Arts. I thank them for their ongoing support. *Building Without Borders* would not have been possible without the contributions of the authors and photographers, all of whom donated their work without compensation. I honor and thank all of you.

My appreciation goes to the steering committee of Builders Without Borders, including Susan Klinker, Alfred von Bachmayr, Melissa Malouf, and especially Catherine Wanek, who all helped shape the book in its earliest iterations. Early critical input from Jean-Louis Bourgeois, Bill Steen, Christel Sumerauer, and Kelly Lerner was crucial in bringing *Building Without Borders* to its current form. Michael Smith brought his critical eye to several chapters. I would like to thank Owen Geiger for reading through the entire book and making numerous suggestions, large and small. Ron Rozewski provided invaluable assistance with the photographs.

Audrey Keating, my copyeditor at New Society Publishers, has played an immense role in the final shaping of *Building Without Borders,* and did the tough work of cutting and shifting where it was needed. As always, New Society's Chris and Judith Plant have been a joy to work with.

I would like to especially acknowledge the patience and support of my wife Rose and daughter Taya, who have persevered with me on what has sometimes seemed a never-ending project. Thank you and I love you.

Lastly I honor the unsung builders around the world, who continue to work without recognition but whose efforts provide the backbone on which a sustainable future humanity will depend. May this book further your work.

Permissions

"The Pallet Truss: A Low-Cost Alternative Roof Structure" by Alfred von Bachmayr appeared originally in *The Last Straw* #38, Summer 2002.

"Speaking the Vernacular: Mud versus Money in Africa, Asia, and the US Southwest" by Jean-Louis Bourgeois is adapted from "Mud Versus Money: Adobe in Africa, Asia, and the U.S. Southwest," Chapter 12 in *Spectacular Vernacular*. © 1996 Jean-Louis Bourgeois.

"Elegant Solutions: The Work of Hassan Fathy" by Simone Swan appeared originally in a slightly different version in *Saudi Aramco World*, July/August 1999.

"Crow Girls' Winning Science Project in not the Last Straw" by Michelle Nijhuis appeared originally in the August 20, 2002 issue of *The Christian Science Monitor* (www.csmonitor.com).

"Seismic Solutions for Straw Bale Construction" by Kelly Lerner is adapted from an article originally published in *The Last Straw* #25, Spring 1999.

"Sustainable Settlements: Rethinking Encampments for Refugees and Displaced Populations" by Cameron Burns first appeared in a different format in *RMI Solutions*, Spring 2002.

"Normal Life after Disasters? Eight Years of Housing Lessons from Marathwada to Gujarat, India" by Alex Salazar originally appeared in *Architecture + Design* (New Delhi) January-February 2002.

Introduction

Joseph F. Kennedy, Co-Founder, Builders Without Borders

Shelter is a human right, yet billions are homeless or living in inadequate conditions due to population pressure, war, or environmental disaster. The housing crisis facing humanity is widely acknowledged, yet few solutions have thus far succeeded in addressing it. Indeed many housing projects proposed to solve the problem are a disaster in themselves: available to few and dependent on energy-intensive industrialized models, the resulting shelters are often inappropriate to climate, culturally inflexible, wasteful, environmentally destructive, and expensive.

What this book offers is a view toward another solution. *Building Without Borders: Sustainable Construction for the Global Village* demonstrates hard-won knowledge about the possibilities and challenges of addressing the global housing crisis in an ecologically sustainable way. It challenges most "development" efforts as being ineffective at best and destructive at worst, but also offers a different path, with a series of Case Studies to illustrate potential solutions. The book takes a holistic approach to the problem of housing provision worldwide and is consistent in its analysis that education and training are the key factors to solving the housing crisis.

The authors featured here champion vernacular traditions, while acknowledging that those traditions must evolve to address changing cultural, ecological, and technological situations. By including a wide variety of authors from many differing countries, academic backgrounds, etc., the book is a true cross-section of the scope of activities currently underway. By avoiding an academic tone, the book becomes accessible to a wider audience of those who want to make a difference, yet remains a useful resource for professionals. *Building Without Borders* provides a grounding in the various issues facing those who hope to address the terrible state of housing worldwide. Although not prescriptive, the book suggests ways to address these situations that will be truly sustainable.

I-1: *It is possible to solve our housing problems using simple materials and human ingenuity, skill, and labor.*

1

The book is composed of Chapters that lay out general issues relating to sustainable construction in a global context. Each Chapter is followed by several Case Studies of particular projects, illustrating themes brought up in that Chapter. Scattered throughout the book are Tech Boxes, detailing technological advances in sustainable construction that can have wide application. Finally several Profiles honor pioneers in this field. The book also includes a Resources section for further research.

Building Without Borders is profusely illustrated with photographs of real projects. Focusing on the hopeful vision of people working together to solve a problem common to humanity, the photos provide a human touch and show the excitement and joy attendant to this most meaningful work — ensuring that everyone has a home.

JOSEPH F. KENNEDY

One note on language: There is no all-encompassing term that describes the nations, countries, or peoples that may, as a group, be faced with housing challenges. Such terms as "Third World," "developing," "underdeveloped," "indigenous," "vernacular," "pre-industrial," and the like are inadequate at best and arrogant at worst. I have, however, decided to keep the terms (including some of the above) as used by the various authors, because a descriptive, all-encompassing term remains elusive. Please forgive any offense this inelegant situation may incur.

The Role of Sustainable Construction in Solving the World Shelter Crisis

Truly sustainable construction supports human dignity, while minimizing negative impacts on the natural environment. The contributors to *Building Without Borders* describe such a way of building — one based on vernacular tradition, on an appropriate use of materials, creative networking, and a human-centered process to create comfortable, decent homes for those most in need. The authors advocate a flexible approach to design, adaptable to available materials and skills and that fit within cultural and social mores. Many authors acknowledge, however, that the addition of minimal modern technology to the timeless wisdom of traditional building techniques can create excellent hybrid structures. These hybrids have greatly improved strength and durability but use locally available, energy-efficient, and Earth friendly materials. Appropriate construction techniques can result in buildings that mitigate environmental damage and, through proper siting and design, save energy by utilizing renewable resources such as sun and wind. Ideally such dwellings would form the cores of sustainable communities where food, water, and waste treatment as well as

I-2: *Children in Northern Ireland being introduced to ancestral construction techniques of building with earth and other building materials, at a workshop given near Belfast.*

economic and cultural opportunities are all collected, grown, or created locally.

In places without decent housing, natural building techniques can be a key component in achieving cheap, comfortable, and easy to build shelter. For example building with straw bales (widely available at low cost) has been revived over the last decade in the US, and has been used with great success to provide low-cost housing in such diverse locales as Mongolia, China, Mexico, Argentina, Belarus, and on Native American lands. Together with earth, stone, and local timber, building with straw can provide shelter much less expensively than can conventional systems (which rely on concrete and steel), while saving 75 percent or more of the energy needed to heat and cool dwellings.

I-3: *Sharing information to solve global problems. Janell Kapoor (far left) has been instrumental in introducing earthbuilding technologies to Thailand to help solve the housing crisis there.*

This is only one example of a range of techniques that serve the goal of reducing a community's dependence on nonrenewable energy and material resources by promoting methods of building that rely on perennially available, renewable resources instead.

But this book is not simply a collection of techniques. The different articles, taken together, describe a potential process by which we can solve the housing crisis. No matter how appropriate the design or materials are, external agencies — whether governments, NGOs or aid organizations — will never solve the problem of homelessness by simply building houses for people who need them. Rather those without homes must be empowered to create them themselves, making use of local skills, native wisdom, and community-centered educational and economic systems. Although sustainable builders have much to offer, only through deep dialog can a truly successful process be created. The goal of this book is to add to this dialog. The ultimate goal is to achieve locally appropriate, ecologically sustainable, affordable, safe, and beautiful homes for all that need them.

I hope that this book inspires and assists those who seek to solve our ongoing housing crisis. By moving past our perceived cultural, political, and racial borders, we can together create a way of housing the people of the world while preserving the Earth, our common home.

What I see with my own eyes is just how long it takes to bring about changes that are lasting. There is a long process of sorting through what works and what doesn't. There are friendships and commitments to develop and maintain. There is the balancing of perceptions from two different cultures about how things should get done. What happens on paper and what actually happens on the ground can be very different. To bring different elements on this Earth a little closer together takes lots of work, and it doesn't happen fast. But all in all, I see progress to somewhere beyond where we are now, and I guess that is what matters most.
Bill Steen

CHAPTER ONE

Shelter and Sustainable Development

Susan Klinker

This chapter provides a foundation for discussions about shelter and sustainable development. It introduces a broad-based look at the current state of affairs, where we have come from, and what is currently evolving in appropriate building technologies and the young natural building revival in the so-called First World. It also introduces some of the overarching social, economic, and political considerations that affect community development and opportunities for the long-term improvement of living conditions.

Population Growth and the Global Housing Crisis

In 1999 the world population surpassed six billion, having doubled in size in less than 40 years. Human populations continue to multiply at exponential rates, with a world population of seven billion anticipated by the year 2010.[2] At the same time the increasingly inequitable distribution of wealth and resources leaves larger percentages of the world's people in poverty. "Globally the 20 percent of the world's people in the highest-income countries account for 86 percent of the total private consumption expenditures — the poorest 20 percent, a minuscule 1.3 percent"[3]

The United Nations Center for Human Settlements (UNCHS) estimates that over 95 percent of the total population growth in the last decade has occurred in developing countries, and anticipates that over 97 percent of continued growth will be concentrated in developing countries in coming years.[4] "In just five years, between 1990 and 1995, the cities of the developing world grew by 263 million people; the equivalent of another Los Angeles or Shanghai forming every three months."[5]

The 1996 UNCHS report identifies that "already more than 600 million people in cities and towns throughout the world are homeless or live in life- or health- threatening situations."[7] "Slums and squatter settlements are now home to an estimated 25 to 30 percent of the urban population in the developing world."[8]

In the context of global population growth and the Earth's finite resources, the way in which human beings are accommodated or sheltered is a major and integral part of the imperative to maintain a global environmental equilibrium. We have entered an era in which no country is isolated and secure from the impacts of the conditions of its neighbors. All countries have a stake in each other's present and future well being. [1]

Of the 15 largest urban agglomerations in 1950, 4 were in developing countries. In 1997, 11 out of 15 were located in developing countries.... The United Nations believes that in 2015, 13 of the 15 largest mega-cities will be in developing countries.... Just 2, Tokyo and New York, will be located in the developed world.[6]

"Since the adoption of the Universal Declaration of Human Rights in 1948 (Article 25), the right to adequate shelter has been recognized as an important component of the right to an adequate standard of living."[9] Despite this it is estimated that over 1 billion people worldwide do not have access to safe shelter and a healthy living environment.[10] The difficulties of ensuring this right continue to raise complex questions:

- How do we define "adequate" shelter?
- Can the basic human right to shelter be sustainably realized within the ecological limits of "Spaceship Earth?"
- Who is responsible for providing shelter?
- How can current processes of shelter development shift to become more socially equitable?

The barriers that limit us from finding real solutions remain rooted in these basic questions. Perhaps by reconsidering contemporary Western values and construction trends from a historical perspective, we can open new possibilities for seeking answers.

Traditional versus Post-Industrial Building Techniques

As long as humankind has been building, people have creatively used materials available in their immediate surroundings and constructed shelter with their own hands. In many ways traditional societies have been the true leaders of sustainable development over time. Inhabitants who are native to a particular place draw their food and water from their immediate surroundings, understanding that their survival requires them to maintain equilibrium with the cycles of life around them. Imbalances eventually cause crises, demanding a reckoning of the scales — usually by population reduction. If not achieved by voluntary migration, rebalancing will likely come about through malnutrition, disease, and either death or a significantly declined birth rate.

Traditional societies have used stone, wood, earth, bamboo, leaves and grasses, animal skins, and even ice to build their settlements, typically building only what is needed to shelter the immediate family and provide for protected communal gatherings. Natural builder Athena Steen describes her family adobe house at Santa Clara Pueblo in northern New Mexico as "... constantly in flux. Walls would go up and then come down. Spaces forever shifted in order to meet changing demands. Not only the

JANELL KAPOOR

1-1: *Monks carrying straw for mud brick construction in Thailand.*

needs of the individual, family, and community had to be met but the larger needs of nature as well. Houses, like people, were allowed to be alive. Structures were built and respected as, over time, they grew, flourished, and eventually died"[11] In Pueblo communities, as in many other cultures, common folk were familiar with basic building techniques and were skilled in the types of repairs necessary to maintain their own homes. Families and whole communities would cooperate to pool the labor necessary for large building projects, passing knowledge from one generation to the next in the process. Consequently communities could be relatively self-reliant in their provision of shelter.

With the advent of the Industrial Revolution, a variety of resources outside the local environment became readily available in many countries at favorable prices. Manufactured products such as kiln-fired bricks, roofing tiles, cement, and metal products could be worked with power-driven tools and easily transported by engine to once remote locations. The new diversity of materials available through the market necessitated increased specialization by tradesmen. No single person could be expected to manage the rising complexity of options available through industrialized building technologies. Although construction speed improved, costs increased and the use of toxic materials in building products became commonplace. Increased regulation by political planning authorities also became necessary to ensure public safety. And little by little, industrialized societies lost their knowledge of local building practices and the ethic of self-reliance in providing shelter.

Compared to pre-industrialized societies, contemporary processes of shelter creation and distribution are alienating. When people simply purchase a home, they are less likely to connect with the local ecology. Furthermore an emphasis on capital outlay limits opportunities for creative expression and community development, and people feel little responsibility for having created the place they inhabit. They are therefore are less likely to view themselves as true stakeholders in the community with an active sense of responsibility for shaping and maintaining its future.

In seeking sustainable solutions to growth that ensure our society's continued survival, we must respect the natural laws of environmental balance. A reconsideration of traditional vernacular architecture can provide a vital link to the further development of viable ecological building solutions to fit a range of urban and rural needs. Accessing the current knowledge of traditional builders is a key component in that link. By concentrating on cooperative efforts and reciprocal learning across cultural, economic, and political boundaries, it may be possible "…to find a responsive and sensitive balance between the still useable skills and wisdom of the past and the sustainable products and inventions of the 20th century."[12]

The Brundtland Commission acknowledged that "indigenous groups are repositories of vast accumulations of traditional knowledge, and that destruction of these societies will lead to the irretrievable loss of generations of wisdom ssociated with managing and living in harmony with complex eco-systems."[13]

Sustainable Development and the Built Environment

Contemporary building technologies of the industrialized world inefficiently use natural and synthetic materials manufactured from resources from across the globe. Rich countries continue to build and "develop" their nations, aggressively consuming disproportionate amounts of the world's natural resources, and projecting images of their standards and ideals for contemporary lifestyles to the far corners of the Earth. Consuming cities import goods of all kinds from remote hinterlands and simultaneously export the majority of their wastes.[14] Current lifestyles and patterns of societal development are unsustainable, impractical, and inappropriate when considered in the broad context of providing safe and equitable living environments for the rapidly increasing numbers of people inhabiting the Earth.

Wolfgang Sachs, who has been a strong proponent of environmental sustainability and a primary critic of over-consumption in the North, notes that humankind has already "consumed as many goods and services since 1950 as the entire previous period of history."[15] Intensified patterns of consumption in industrialized countries can be seen as the root cause of sustainability problems today.

When considering near-future projections of population growth and the impenetrable boundaries of "Spaceship Earth," it seems clear that more sustainable shifts in technology development are necessary. In 1983 the United Nations General Assembly appointed the World Commission on Environment and Development to "propose long-term environmental strategies for achieving sustainable development." The commission defined sustainability as "those paths of social, economic, and political progress that meet the needs of the present without compromising the ability of future generations to meet their own needs."[17]

The Natural Building Revival

The natural building revival currently gaining attention throughout America and Europe begins to offer some significant alternatives to dealing with the problem of sustainability and our built environment. Advances in natural building provide solutions that meet and often exceed consumer expectations in the levels of comfort, esthetics, and quality demanded of modern buildings. Natural building technologies can be applicable in nearly all geographic regions and climates on the Earth.

In addition to straw bale construction, the use of adobe, cob, rammed earth, and straw-clay are growing in popularity. Alternative structures are also being built of recycled materials such as tires, aluminum cans, paper fiber, and bottles combined with cement or earth-binding agents. A growing group of owner-builders in the US has

"If all countries successfully followed the industrial example, five or six planets would be needed to serve as mines and waste dumps. It is obvious that 'advanced' societies are no model at all; rather they are most likely to be seen in the end as an aberration in the course of history" in other words, "a blunder of planetary proportions."[16]

explored the advantages of low-tech natural building and has created cost-effective and comfortable housing that was built with their own hands. Luxury natural homes have also become increasingly trendy for high-end, custom-built homes in the US. The factors influencing conscious consumer choices are numerous, including: environmental sensitivity, beauty and hand-hewn esthetic qualities, comfort, affordability, health concerns (chemical sensitivity), a commitment to the principles of non-consumerism, or simply the desire to live a lifestyle in closer harmony with nature.

1-2: *Centers of innovation in sustainable design and construction have sprung up all over the world, such as in the community of Ecoaldea Huehuecoyotl near Tepoztlan in Mexico.*

Popular acceptance of alternative sustainable building techniques has grown, especially for straw bale construction. Media coverage in the US, including profiles on *Good Morning America* and *This Old House*; in journals such as *Architectural Digest*, *Fine Homebuilding*, *Natural Home Magazine*, and *Designer/ Builder*; and in newspapers such as the *Wall Street Journal*, *New York Times* and *Los Angeles Times,* has helped to increase awareness and public interest. International agencies, such as UNCHS, also acknowledge the potential for natural building technologies to resolve housing shortages worldwide.

However, in many cases, appropriate low-tech natural building alternatives are rejected in new housing programs because they do not seem to fit with contemporary Western ideals and standards. Barriers can be both social and institutional. At a social level, families want their housing investment to reflect a positive image and increased social status. These desires are often closely tied with perceptions of modernization and a rejection of traditional ways that may be thought backward. At the institutional level, many government and financial institutions fail to acknowledge the validity of traditional building alternatives for new construction. They may also associate the use of traditional building methods with a lack of progress. Capital investment into these types of projects may also be considered high risk or even a waste of financial resources if market values for natural built homes are not assured long-term.

Therefore, as we enter the new millennium as a global community, the playing field for the development of more sophisticated and appropriate building technologies is equally grounded in both the so-called industrialized First World and the less industrialized developing world. We must dissipate prejudices and misconceptions about ecological building technologies through projects that demonstrate affordability,

KELLY LERNER

1-3: *Traditional architecture, as exemplified by this stone house in China, is unfortunately under attack by culturally destructive political and economic interests and the adoption of a symbology that equates tradition with backwardness.*

longevity, beauty, and overall excellence in design. We must write up case studies that document lessons learned from individual projects so that we can share meaningful information and reveal trends that may be applicable in a diversity of situations, regardless of cultural or socio-economic context. Development programs that take an approach of reciprocal learning and shared experience with traditional builders and that seek to effect change in both realms will ultimately prove to be more effective than those aimed at technology transfer.

Empowerment and the Participatory Development Process

The word "development" implies favorable change from worse to better.[18] Yet all too often, well-intentioned development projects have had disastrous effects on local people, including the disruption of social and political environments, the invalidation of traditional values and means of production, and the creation of new desires and dependencies. Therefore we must carefully consider any involvement by outsiders that is intended to effect change on a community level.

Beyond basic material gain and the physical outcomes anticipated by a project, it is important to consider how any proposed project will empower local people. Who will be empowered and with what? Will everyone benefit equally, or only those who meet specific criteria? Will participants gain knowledge, skills, status, or confidence?

It is also important to evaluate what kind of relationships the project will develop. Will there be long- or short-term relationships among team members? Will information exchange be reciprocal? Will the proposed process increase independence or improve local communal relations? Most importantly, intervention workers must maintain a sense of humility and respect for the process they are engaged in, recognizing what is changed through the project process and why. Sensitivity to the subtle responses of the community and a sincere, conscious effort to monitor project impacts at every level will help to keep initiatives on target.

When community members are involved in establishing priorities and in decision making, they are invited to become true stakeholders in a project. Their grassroots involvement helps to expand their resources for future problem solving while allowing them to experience the immediate results of their decisions as the project proceeds. A well-planned community design and planning process can not only improve physical conditions, but can also help to build pride and raise a community's sense of responsibility for creating and maintaining a healthy environment. While working together, residents from varying social and ethnic backgrounds often find new understanding of each other and create new common ground for moving forward as supportive neighbors. Strengthened relationships and partnerships within the community also reinforce local capacities and help ensure the long-term sustainability and health of the community once planners or supporting NGOs withdraw upon project completion.

If efficiency in achieving physical outcomes is the primary goal of a development agency, then grassroots participatory planning will probably not be the most effective approach.

The more people involved in the decision-making process, the longer it takes to get things done. Therefore, when working in a development context, agencies must carefully consider their goals and available resources and make conscious decisions about potential opportunities for education, networking, and capacity building at the local community level.[19] These decisions may lead to greater long-term impacts and sustainability on many levels. In any case, community participation and the project process "should not bring simply the direct material benefits…but a transformation of consciousness and self-perception."[20]

IVAN YAHOLNITSKY

1-4: *This house in Lesotho integrates traditional materials and forms with appropriate technologies such as photovoltaics, water catchment, and passive daylighting to create an improved vernacular architecture.*

Social and Economic Considerations and Sustainable Development

Planners and development professionals often misunderstand the hierarchy of basic needs and priorities of the poorest. If a family is hungry, the need to provide food is of far greater importance than is shelter, especially in warm dry climates where it is comfortable to be outside.[21] Employment security, food, healthcare, and opportunities for education are often seen as priorities over improved shelter.[22] When families are forced to allocate too high a proportion of their income to housing, they are left destabilized in other areas of their lives.[23] Therefore, intervention programs aimed at sustainable development must also consider the broader issues of economic development, social justice, and ecological responsibility.[24] These issues go hand in hand in creating physical and social environments where people are empowered and can gain access to resources that enable them to more actively engage in creating their own futures and livelihoods.[25]

Many other aspects of human settlements must also be addressed. Due to increasing land values, for example, the majority of low-income settlements are located on the periphery of urban centers, making access to employment costly and inconvenient. The distribution of such basic services as roads, drainage systems, and electrical/telephone service has an impact on the very local business development that can be so essential for economic prosperity. Furthermore, a lack of sanitation services such as water, sewer, and garbage collection can pose serious health risks from biological hazards spread by rodents and insects.[26] And poor air quality and inadequate ventilation can aggravate bronchial problems or cause permanent lung damage. On the other hand, convenient access to public services such as education, healthcare, and public transportation significantly improves people's potential for employment, equal opportunity, and overall quality of life. Consideration of all of these factors is important in addressing balanced community development and the long-term sustainability of a settlement.

Many self-help projects are specifically designed to advocate incremental building. Rather than requiring a large initial loan, incremental construction allows a household to upgrade or add more space as financial resources become available. The quality of the house may be much lower during early phases of a project, but the overall security of the family is enhanced. For example a typical "sites and services" project will provide water, sewer, and electrical service to a specific plot of land. Many also include a simple foundation and a concrete slab floor. Beyond these basic sites and services, families are free to build as they wish, within their own time frame. To support such projects, banks may need to implement policy measures and incentives to free up small loans for construction and adjust the types of finance available to low-income families.[27] Other physical planning decisions may also affect community economic opportunities, such as making allowances for income-generating activities within the home and providing customers with easy access to home-based commercial enterprises.

An intentional development of local cottage industries to support the building trades (based on the use of rapidly renewable local resources) can have a significant impact on a community's ability to afford housing improvements. Low-tech construction methods typically emphasize the use of labor-intensive rather than capital-intensive technologies and are readily applicable in low-skilled, self-help housing development schemes. In addition to common materials such as wood, stone, earth, and bamboo; agricultural or industrial by-products such as straw, volcanic pumice stone, rice or coconut husks, and lime may also be used to develop building products.[28] Natural resources may need to be harvested sustainably in ways that do not contribute to local environmental degradation. Such an approach not only supports new economic development initiatives but also fosters self-reliance and sustainability in the provision of shelter. And it has considerable potential for supporting the rapid expansion of the local stock of affordable housing.

Gender Perspectives in Housing

As many as 30 to 40 percent of the world's urban households are currently headed by women.[30] This clearly calls for a gender-sensitive approach to planning individual housing units and communal facilities. Many women in developing countries operate a shop or food store from their home, or participate in cottage industries (such as sewing or manufacturing piecework), while also providing primary childcare. Thus planners must address safety issues from a women's perspective. Communal wells and latrines or sanitation facilities must properly provide for appropriate modesty and security relative to each specific cultural context. Spatial consideration of these kinds of domestic activities is critical to creating opportunities for women to better provide for themselves and their families.

Although women in developing countries are often highly active in the physical construction of their homes, their ability to fully participate in a self-help building program may be subject to cultural values and mores, as well as to physical limitations due to age or health. In addition the triple role of female heads of households (as mother, homemaker, and primary source of economic support) may impose severe time limitations on their ability to build a home on their own.[31] In planning for self-built housing schemes, it is important to address the unique factors which may limit women's ability to fully participate in the program and to devise ways in which women may better access equal benefit.

Policy, Affordability and Building Codes

Definitions of adequate shelter vary widely in different nations, cultures, and social circumstances. What is perceived to be very adequate to members of a traditional society living in a rural agricultural area may be considered highly inadequate to those accustomed to Western-style urban dwellings. Through his work in Peru in the 1960s, John Turner presented new ways of thinking about housing. He introduced the concept of viewing housing as a verb, recognizing "what housing does" rather than "what housing is." Turner's view stresses the "use value" of a house (as determined by how well it serves the needs of its occupants) rather than its "market value" (which is defined in dollars and cents). A "use value" perspective encourages alternative forms

The criteria for selection of a technology should include the affordability of its initial cost, the ability of local labor to utilize and maintain it, and the viability of adapting it to local needs. [29]

PAULINA WOJCIECHOWSKA

1-5: The revival of timeless building techniques (as exemplified by this earth plastering of a straw bale house in Poland) is a sign of hope that we are rediscovering ways to shelter people without destroying the planet.

of housing, variety in the ways people attain shelter, and the use of the housing process itself as a "vehicle for the fulfillment of users' lives and hopes."[32]

Balancing economic extremes, and establishing appropriate standards for health and life-safety without infringing on people's rights or abilities to access shelter poses challenges for policymakers. If housing policies and building regulations are intended to support the fulfillment of the basic human right to adequate shelter, then they need to be flexible enough to accommodate the full range of people's needs, desires, and abilities in attaining it.

Unrealistic standards of construction and impractical code requirements can limit the availability of housing for the middle class as well as for the poor. Many building codes are modeled after those used in industrialized countries and often favor high-tech, import-based materials that can be scarce and expensive in developing countries. Codes sometimes even prohibit the use of traditional building methods. Although these types of requirements may be inappropriate, many governments are unwilling to change their defined standards because they fear criticism that they are reducing standards and sanctioning the development of low-quality housing and, therefore, the further growth of slums. When building codes are out of balance with local needs, circumstances can breed corruption as people find it impossible to build according to required standards. They may bribe officials for necessary building approvals or proceed knowing that their housing is considered illegal.[33]

The adoption of distinctive codes and construction standards for traditional or alternative building techniques could significantly affect the availability of safe and affordable housing worldwide. In many cases ecologically benign or recycled materials are perfectly adequate for simple construction needs, and have tremendous potential for improving the quality of existing housing when used in the right way.[34] What is still lacking in most building codes are adequate guidelines for artisans and professionals regarding the safe use of indigenous materials and low-tech appropriate technologies, and allowances for the incorporation of feasible innovative technologies as they

According to the 1988 UNCHS Global Strategy for Shelter (GSS), adequate shelter means "adequate privacy, sufficient space and security, adequate lighting and ventilation, adequate infrastructure in a location with adequate access to employment, and basic services at a price which is affordable to the user."[36] Affordability, as defined by UNCHS standards, usually falls between 15 and 25 percent of household income.[37]

emerge. Ongoing public demand, activism, and lobbying efforts by those who have demonstrated an ability to produce sustainable shelter effectively are critical to continuing the reform process.[35]

Toward Sustainability

When dealing in projects oriented toward sustainable development, it is important to consider the bigger picture. Tough questions asked during project development can help ground the project at a more meaningful level. For instance: How can the project process and communication serve to create a sustained and positive impact in the community? How will decision-making processes affect community relations and offer new opportunities? Are team members open to the experience and willing to bend as new aspects of the work unfold? Are the unique needs of women considered? Will the project increase economic opportunity? Will it increase respect for local environmental resources? Are project discussions oriented toward reciprocal learning? And how can this project link with the larger community (local and national officials as well as the global community), help to increase awareness, and share information across cultural and political boundaries? These questions focus on only some of the considerations that will help to integrate a project within broad-based social and environmental agendas. The hope is that by asking them, we may collectively progress more rapidly and effectively toward a more sustainable future and more equitable contions of shelter for all.

Ismail Seregeldin, author of *The Architecture of Empowerment*, expresses the challenge before us so well:

> Poverty and environmental degradation go hand in hand.... The creation of humane environments and livable cities is within our grasp. No revolutionary new technology is required; we know the methods and have seen them work. The will to implement what we know and the determination to succeed in doing so is what is needed.... This is the challenge before us. [39]

More than 100 world leaders participated in The 1992 Rio Earth Summit. The publication of their proceedings, Agenda 21, includes the following recommendations for the management of human settlements (partial list):

- The use of local and indigenous building sources
- Incentives to promote the continuation of traditional techniques and self-help strategies
- Improved construction materials, techniques, and training in recognition of the inequitable toll that natural disasters take on developing countries
- The regulation of energy-efficient design principles
- The use of labor-intensive rather than energy-intensive construction techniques
- The appropriate restructuring of credit institutions
- International information exchange
- The recycling and reuse of building materials
- Decentralization of the construction industry, through encouraging smaller firms
- The use of "clean technologies." [38]

Case Study:
Tlholego Development Project — A Sustainable Ecovillage in Rural South Africa

Paul Cohen

For diverse reasons — separation policies during apartheid, the proliferation of large farm systems, and rapid industrialization and subsequent migration to urban areas — many South Africans have become alienated from formerly thriving rural areas in recent decades. Policies of mobile labor and lack of land tenure have disrupted family structures, forcing workers to leave their partners, dependent children, and elders in depressed rural villages.

PAUL COHEN

1-6: *A township in Cape Town. Despite new political freedoms, most people in South Africa still live in substandard housing.*

Despite newfound political freedom in recent years, many South Africans still face significant obstacles to achieving economic freedom. In addition to the challenges of entering a highly competitive global economy, the millions of South Africans living in poverty have an overwhelming need for land tenure, jobs, housing, and food. Efforts to resettle these residents constitute perhaps the biggest challenge facing the new South Africa.

In response to South Africa's situation, low-cost housing solutions from around the world have flooded in as foreign and domestic companies vie for space in a potentially lucrative market. Yet the resulting hastily built settlements rarely meet residents' basic needs and leave little possibility for community growth.

In 1991 I formed a nonprofit organization to initiate the Tlholego Development Project (TDP), after purchasing a 300-acre (120-hectare) degraded farm near the city of Rustenburg in the Northwest Province of South Africa. Here we have created a pattern of village development that introduces new forms of community organization for both for rural and urban areas of South Africa, using ecological principles for guidance. We knew that recreating sustainable rural villages would help to balance migration from the countryside to overcrowded cities.

TDP, the resulting model, demonstrates and teaches sustainable approaches to land use, housing, food security, and village development. "Tlholego" is a Tswana word meaning "creating from nature." Through working with leading professionals from South Africa and around the world, TDP has facilitated many of South Africa's primary training programs in ecovillage development, sustainable building technologies, and permaculture design. Practical demonstration of village settlement and rural economic development integrates traditional African design, modern technology, and lessons learned from around the world, giving hope and creating positive aspirations for communities living in rural poverty.

Tlholego is comprised of three core components. The Tlholego Educational Institute researches, develops training programs, and provides a nurturing environment for growth on a human and economic level. The Tlholego Residential Village employs ecovillage design, permaculture, and natural building to create a residential model for land tenure, sustainable housing, and local economy. And the Tshedimosong (meaning "place of enlightenment") Farm School, a primary-secondary school provides basic education to 120 children from the surrounding farming community.

Addressing South Africa's Housing Crisis

According to USAID (1996) an estimated 48 percent of South Africans reside in dwellings that are deemed inadequate.[1] South Africa's marginalized urban majority live in uncomfortable corrugated metal shacks, where they sweat in summer and shiver in winter. Further, during winter months, families are exposed to high levels of indoor air pollution, which result from the combustion of fuels burned for warmth.

The national government's housing subsidy provides basic sites and services in the majority of cases, with often a simple roof on four poles as the house. Most low-cost housing systems achieve cost efficiency by using standard house design and standardized minimum cost materials. Governments and housing authorities generally accept these techniques as the best methods for producing low-cost housing en masse. Large building companies also like this approach because it often results in maximum profit through repetition.

The houses produced, however, are usually very low in quality — particularly regarding thermal, environmental, and esthetic characteristics. The thermal problems created by these designs can lead to high costs for heating and cooling. End users are usually not consulted in the design process or given alternatives, and they often experience low levels of satisfaction with the houses (often as a result of the houses being too small). Future additions at a later date are not planned for and can be difficult to bring about.

1-7: *Traditional building skills such as round pole building and thatching are demonstrated at the Tlholego Development Project.*

In owner-built sustainable building strategies, I found an approach that could provide better housing solutions than could standardized housing, and for a similar or lower cost. In 1994 the founding residents of Tlholego began their practical training in sustainable building technologies. I invited natural builder and architect Joseph Kennedy to develop some prototype structures and help create a master plan for the site. Over the following two years, owner-builders constructed a series of experimental buildings. Some were constructed using traditional 2000-year-old Tswana designs that made use of earth and thatch. Others were constructed from large earth-filled bags and had fired-brick dome roofs. Locally available and recycled materials were used for foundations, walls, floors, and roofs.

In 1996, after accumulating experience with owner-builder methods, Tlholego began working in partnership with Brian Woodward (an Australian unfired-mudbrick building expert) to develop the Tlholego Building System (TBS), a sustainable housing system for South Africa. In this approach we first determined what we consider the minimum size and quality requirement necessary for a family home in South Africa. Then we looked for ways to construct this minimum house for the money available. Our goal was to create a flexible, owner-built, low-cost, high quality housing system that addressed the serious shortcomings of typical low-cost housing construction in South Africa, while considering environmental and resource issues not usually considered.

The housing strategy combined the principles of sustainable building systems with natural waste treatment and the permaculture approach for designing food self-reliance. This strategy includes using modern techniques of unburned mudbrick; passive solar design; appropriate technologies of rainwater collection, composting toilets, greywater irrigation (water from sinks, shower, etc., excluding toilets); and solar water heating.

Our first prototype house in 1996 was a four-room, 450-square-foot (42-square-meter) family home for Tlholego site manager Fanki Mokgokolo. It featured passive solar design with shower, laundry, and kitchen area, along with dampproofing, termite protection, insect screening, high quality surface finishes, on-site waste treatment, and electricity in every room. The price limit we set was R8,000 (US$1,070). From the beginning we chose owner-building because of the cost savings it achieves. By selecting labor-intensive but low-cost techniques, we reduced materials costs even further.

TDP's houses are designed to last far longer than those built of conventional materials. By establishing permanent food systems in the immediate vicinity of the home, we also create additional esthetic and functional outdoor living environments far larger than the internal dimensions of the house — another asset that will appreciate in value.

Our approach to housing can be widely applied, with plenty of flexibility to allow for designs to differ according to one's personal criteria. If the cost implications of alternatives are clearly spelled out, then the owner-builder can make an informed choice. The choice may be between different materials (based on cost or availability) or between quality and quantity (a larger, lower quality house or a smaller, higher quality house). The flexibility of this approach also facilitates staged construction and/or extensions at a later date.

Ecological Design and Construction

The building system most commonly adopted at Tlholego is a locally manufactured mudbrick wall set on a concrete, block, or stone foundation. The walls are well attached to a lightweight, insulated timber and metal roof. The walls are rubbed down with water to reduce cracks and provide a pleasing texture and are then coated with linseed oil and turpentine for weather resistance. We use passive solar techniques such as solar orientation, thermal mass, and proper overhangs (see "Sustainable Building As Appropriate Technology"). Shade extensions are attached to the most vulnerable sides on the house, and vines or trees can be grown to create protection from driving rains.

PAUL COHEN

1-8: The local clay earth is used at Tlholego to create modern looking, comfortable, and inexpensive dwellings.

Natural Waste Management

The houses at Tlholego integrate on-site waste management for safely managing human wastes. The main system developed at Tlholego is a composting toilet that is low-cost and easy to construct; it uses only the most basic available building materials, has no moving parts, is robust, and has wide-scale application.

Water Harvesting

The harvesting of water is a key element of any sustainable housing and land use system. The more water a household is able to harvest the greater the level of food and water security. This strategy can be applied equally well at the household level as at the village or watershed level.

Ideally the first level of water catchment takes place on the rooftop through a gutter system that empties into a storage tank, which is then available as a high quality

PAUL COHEN

water source for drinking or irrigation. If there is insufficient funding for water tanks and guttering in the initial stages of construction, water flowing off the roof during the rainy season can be directed along the ground into tree plantings, or to food gardens around the house. Excess water can be directed into below-ground storage and used for irrigation during the dry times of the year.

Wastewater is another available source of water at the household level. Our buildings incorporate greywater filtration, which cleans wastewater so that bathroom and kitchen wastewater can be used for irrigation purposes.

1-9: Buildings are integrated with landscape. In this case water channeled off a barn roof is used to create a pond for wildlife habitat, food security, and beauty.

Permaculture Food Security

There is a natural integration between our buildings and the "food security" gardens that have been established in the immediate vicinity of the houses. Permaculture, now in wide use around the world, is a design system for human landscapes developed by Bill Mollison and David Holmgren. It integrates traditional systems of knowledge with modern science and common sense to provide a complete food production system incorporating annual and perennial plants, small animals, and useful trees. It relies on low external energy inputs. The beauty of this system is that, in addition to establishing a valuable source of healthy food, it increases the size of a family homestead's living environment and adds a tangible dimension of quality to a rural or urban lifestyle.

Accomplishments

One of the important accomplishments thus far has been the sustained transfer of building skills that Tlholego residents have experienced since 1994. The techniques have stood up well to residents' expectations, and Tlholego is now proud of having established the first competent building team in Southern Africa capable of spreading these systems within the broader community.

In February 2000 our village model was chosen by the National Department of Housing (DOH) as the most appropriate model to represent South Africa at the Africa "Solutions Towards Sustainable Development" conference in March 2000.

Education and Training

Over the past 12 years Tlholego has accumulated considerable knowledge about sustainable building technologies. This experience has grown through hundreds of hours of shared learning between professional architect/builders from around the world and Tlholego community members, most of whom have come from a rural farm worker background with a minimal skills base. The result is that we have now established a small but competent building team capable of building new houses at Tlholego and of training other communities and people from around the world. TDP is an internationally recognized reference point and demonstration center for sustainable living, and can offer education and training in sustainable building on several different levels.

Goals for the Future

Our short-term goal is to raise financial support in order to replace more of the substandard housing units that many of our community members occupy. New construction will create an ideal and much-needed opportunity for people from the local community, potential trainers, and interns and volunteers from around the world to gain practical experience in building techniques, as well as to learn about sustainable rural settlements.

We also plan to expand our educational activities over the following few years. Thus far I have largely directed activities, assisted by team of consultants and interns. In an effort to empower more local people with an interest in these ideas, the Tlholego Educational Center and bioreserve is being restructured to allow new partnerships to form in the areas of education, ecotourism, and sustainable agriculture. As ecotourism opportunities grow within the region, Tlholego aims to build upon its initial foundation and continue to establish itself as a unique venue, providing an authentic African experience and connecting the sustainable knowledge of the past and the future as a gift in the present.

Lessons Learned

The innovations of Tlholego are unique to the South African context, but many of the lessons learned from our experience can be transferred to projects having similar goals in other parts of the world. Among these lessons are:

- Systems from other situations can't be transferred wholesale; they need to respond to local conditions.
- Visiting consultants should defer to local experience. It is important for these visitors not to get too attached to romantic notions developed out of context, or to their "expert" opinion. Tlholego suffered from some well-intentioned but ill-prepared or inconsiderate experts in its development.
- Do not underestimate the power of outside influence. Even in the most isolated rural areas, people have a sense of what is out there in the world

through media exposure of various kinds. This is especially true in a relatively developed country like South Africa.

- Manufactured materials are sometimes more sustainable than the unsustainable harvest of natural materials.
- Political/financial considerations can become the make-or-break factors of a project's success.
- It is important to revisit projects to see if they stand the test of time. Early experiments in plastering proved to be unsuccessful, while techniques that were successful elsewhere were not as successful at Tlholego. Certain styles of architecture were not as readily adopted as were others. These lessons can be used to create a robust new system.
- The people in South African rural situations like Tlholego have suffered from decades of abuse, little education, and dysfunctional family situations. It is unrealistic to expect leaders capable of running projects as sophisticated as Tlholego to emerge from such a context in the near term. It is a long-term process that must be nurtured to allow the abundance of human potential to emerge, evolve, and become part of the "social capital" so valuable to the smooth running and growth of a sustainable village settlement.
- Beauty is important. One of the great strengths of the work at Tlholego is its esthetic beauty. This power to inspire is intangible but remarked upon time and time again by visitors to Tlholego.

Tlholego Ecovillage has shown that at our relatively small scale of activity, it is quite possible to harmonize the basic elements of sustainable development — land, housing, health, food, water and sanitation, education, and employment — into an integrated system that transforms current patterns of consumption and poverty. If, like a tree born out of a relatively small seed, which is driven by still smaller genetic material, our projects ensure that core systems are designed sustainably, it is natural that anything that grows from this core will also be inherently sustainable. Thus, in this way, like a powerful biological agent, we will be able to create a world capable of supporting our unfolding human potential.

Case Study:
Straw Bale Construction in Anapra, Mexico

Alfred von Bachmayr

In 2000 Builders Without Borders (BWB) helped coordinate one of its first projects in Anapra, Mexico. Anapra is a *colonia* (informal settlement) on the north side of Ciudad Juarez, situated within sight of the United States/Mexico border. A typical squatter settlement, Anapra has grown into a large community. And as is common in such *colonias*, it is difficult for the residents to produce adequate housing due to lack of financial resources. However the people are incredibly resourceful and creative in the ways they build housing for themselves.

The climate of this arid desert region is severe — very hot and dry in the summer and cold and windy in the winter. Houses are commonly built of concrete or concrete blocks with lightly framed uninsulated roofs or pallets covered with tarpaper. These houses, however, do not adequately protect the families from the severe climate, and people's health is compromised.

BWB assisted in plastering Casa Ameas, a straw bale community center in Anapra. The center had been built by Annunciation House (a Catholic organization that cares for challenged individuals and families on both sides of the border) as part of their efforts to address the severe shortage of habitable structures in the region. The plastering workshop itself was a great success. And it generated additional opportunities for BWB to help with the border housing crisis.

1-10: *Economic policies that make rural livelihoods difficult or impossible, together with the relocation of factories to take advantage of "cheap labor" have led to the proliferation of informal settlements, known as colonias, along the US/Mexico border.*

A workshop volunteer introduced us to Jose Luis Rocha, one of the lead builders of Casa Ameas. Jose Luis needed a house for his family of six, as his previous dwelling had burned down. BWB quickly realized that working with Jose Luis would give us a chance to develop a model straw bale home that could be easily and cheaply replicated throughout the region.

The Jose Luis Rocha House

BWB began consulting with Jose Luis in the summer of 2001, at which time he supplied us with ideas for his house. I acted as project architect for BWB and turned his ideas into drawings for construction. We chose straw bale construction, partly because Jose Luis had been involved in building Casa Ameas, but also because he liked the technique for its simplicity, low cost, insulation value, and esthetics. I proposed that the roof structure be built of trusses constructed of disassembled pallets that could be reassembled in a calculated configuration to be sufficiently strong to bear the roof loads (see "The Pallet Truss"). We prototyped the pallet system at a BWB training session in New Mexico, together with the straw bale wall system, prior to actual construction.

Jose Luis built the foundation and acquired straw bales, pallets for the roof trusses, and miscellaneous building materials prior to the arrival of 15 BWB volunteers in October 2001. Project managers Alfred von Bachmayr and Catherine Wanek led the volunteers, including several who had taken the BWB training session in New Mexico. During approximately ten days of hard work under challenging conditions, the straw bale walls, roof trusses, and roof insulation were installed for half of the building. (BWB initially committed to build only half of the 1200-square-foot [110-square-meter] design that Jose Luis wanted.) BWB volunteers also constructed the trusses for the other half of the building.

A few volunteers stayed to help Jose Luis build the other half of the house. In subsequent work efforts, the entire structure was completed and roofed. And over the following months, Jose Luis and his family continued to finish the building as resources and time were available. In February 2002 BWB initiated another work effort — 14 volunteers helped complete the interior and exterior plastering — and raised money to purchase windows and doors.

Casas de La Cruz

During the construction of Jose Luis's house, a local missionary group called Casas de la Cruz (CDLC) came to the construction site and expressed interest in straw bale construction for their own building program. (Over the previous 14 years, CLDC had been building approximately four houses a year out of concrete block for local families. They had a family selection committee, consisting of several women who were now living in homes built by the organization.) Having seen Jose Luis's house during its construction, as well as Casa Ameas, the group asked BWB volunteers if they would be willing to help build a straw bale house financed by CDLC. The request supported BWB's mission of spreading sustainable building technologies with local partners, so BWB agreed to assist them and proceeded with the design of a house. A family (consisting of a single mother with two children) was selected and the house sited on its lot.

In February of 2002 two BWB project managers (Melissa Malouf and I) directed the construction of the 15-by-30-foot (4.6-by-9-meter) load-bearing, passive solar straw bale building. (To help finance BWB's involvement, two paying interns participated in the construction and were given building training by the project managers.) BWB members, leaders, and CLDC volunteers constructed roof trusses of used pallets and set them on a 2-by-4-inch box beam. We insulated the ceiling with straw bale "flakes" approximately 12 inches (30 centimeters) deep, which were sprayed with liquefied clay on both upper and lower sides to provide fire protection. We installed metal roofing on trusses that had approximately 24-inch (60-centimeter) overhangs. And we poured a 2-1/2-inch-thick (6-centimeter-thick) concrete floor. Locally made windows and doors were installed, and we plastered the building with mud both inside and out.

1-11: *Jose Luis Rocha stands proudly in front of his house built (with the help of Builders Without Borders members) of locally available materials such as straw, recycled concrete, and wooden pallets.*

Lessons Learned

Jose Luis's House

First, because of its size, Jose Luis's house could not be finished in a reasonable amount of time, leaving completion to the family. This proved to be a real challenge for them (because of limited financial resources), and the house was still not being lived in a year after construction began. As a result we have decided that BWB should only take on projects that can be completed by the organization and its partners in a reasonable amount of time. In Jose Luis's case, BWB could have committed to — and finished! — a structure half the size of the one built. Then, if another had been needed to serve the family, it could have been executed as a subsequent project.

Second Jose Luis chose to plaster the outside of his house with cement stucco, an expensive choice that proved to be beyond his financial resources. Earthen plaster would have been less expensive, more user friendly, and quicker to apply. In future projects we need to discuss the pros and cons of plasters and other materials with prospective owners when the buildings are conceived.

On the positive side Jose Luis became the most knowledgeable local person in straw bale construction and was hired as the maestro to work on the CDLC house. He worked hard to learn how to build pallet roof trusses, and it is hoped this can become a micro-enterprise for him. BWB will lead another work group to assist the family in their completion efforts.

1-12: *Caption: Local women plaster the interior of a new straw bale home in Anapra, Mexico.*

The CDLC House

The small residence built with CDLC for Maria and her two children was completed in two weeks for approximately US$3,700. The cost of future houses could be significantly reduced through incorporating efficiencies developed during construction of the first house and through more effective purchasing of materials. A larger machine to chop straw for plastering, for example, could save a great deal of time. The pneumatic texture-spraying gun we used for coating the insulation and first coat of plaster proved to be a huge time saver and will be used in future projects. It took several tries, however, to develop a mud plaster mix that wouldn't crack. In future projects a water-resistant final layer should be considered and developed.

The community was reluctant to accept the straw bale building at first. The most resistance came from the committee of women ordinarily involved in the construction of new houses: they felt displaced due to lack of familiarity with the new construction technique. Maria, however, was elated with the building and says she is very comfortable even on the hottest days. And, apparently, as more people experience the building, it is gaining acceptance. BWB recognized from this experience just how important the inter-personal aspects of a project are.

CDLC intends to build four more straw bale buildings this year and has again asked BWB to help with design revision and to head the construction process. Their commit-tee is enthusiastic about finding consistent sources for straw — one of our greatest challenges in the first building.

Future Developments in Anapra

Future projects will involve integration of water service, water distillation, and some human waste disposal systems (either composting or a wicking bed system) with the straw bale and pallet truss house model. We hope to involve more local people in the projects and train them in the construction of foundations, trusses, and waste systems. Through this local involvement, we hope that the program can become more and more locally based.

Through partnership with CDLC and the infrastructure they have established in the community, we now have the opportunity to produce many more houses each year. We feel confident that this effort can grow to make a significant difference to the housing problem in the Anapra region.

THE PALLET TRUSS:
A LOW-COST ALTERNATIVE ROOF STRUCTURE

Alfred von Bachmayr

In my work in the border area around El Paso/Cuidad Juarez, I have been amazed at the number of uses people have devised for wooden shipping pallets. I have seen everything from full houses to an incredible variety of fencing designs, furniture, and even kids' toys. I realized that the low cost and availability of pallets makes them a reliable and versatile resource. At the same time I was struggling with what seems to be a universal challenge — finding low-cost structural members for roofs that would allow for thick roof insulation. The typically available 2-by-4-inch or 2-by-6-inch rafters work structurally in some areas with low live loads, but the joist depth does not allow for sufficient roof insulation. The idea of a truss made from pallet parts seemed like it could be the answer to these challenges.

With the help of some local carpenters, I began trying different truss configurations and joint connections. We looked for pallets that had full-length 2-by-4-inch members and avoided pallets with slots cut out to allow the forklifts to access them from all sides. After several prototypes we developed a simple way to construct trusses that also proved to be structurally sound. I had a structural engineer model the truss on a computer and give me the required number of nails for each joint. We also decided to use glue at each joint to achieve the extra capability needed, and the nails were put in from both sides to tighten the joints. Any configuration of truss could be developed, assuming the length of the pallet parts was not exceeded in the design.

The first challenge was to find a way to disassemble the pallets. Pulling the nails was almost impossible without destroying the parts. However we found that by cutting the nails with a reciprocating saw, using 8- to 9-inch (20- to 30-centimeter) cutting blades, the pallets would come apart quickly with no damage to the parts

To standardize the construction of the trusses so that they were dimensionally stable and consistent in strength, we built a jig out of 2-by-4-inch rails nailed to plywood or oriented strand board (OSB) that mirrored the shape of the outline of the trusses. (We drew all the parts on the jig so everyone knew exactly where all the parts went to construct the trusses.) We cut 12-inch-long (30-centimeter-long) gusset plates and the diagonal braces from the slats, and also squared the ends of the struts on a chop saw. Then, using the jig, we began building the top and bottom chords. We put 12-inch (30 - centimeter) gussets in the jig at each joint of the 2-by-4-inch struts, then glued the gussets and put down the precut 2-by-4-inch struts/chords and clamped them to the rails of the jig. Then we put glue on the contact point of the chords and finally the other gussets on top. We nailed through all three pieces with six 8d nails on each side of the butt joint. We then flipped the chord over, reclamped, and put the same number of nails on the opposite side. We did this for all the chords needed for the truss.

With all the chords constructed, and using the same jig, we put down the bottom diagonal braces. These were glued on top where they contacted the chords. We then clamped and glued on the top and bottom chords, and finally added the top diagonal braces. We then nailed through all three pieces with 8d nails, three per side. As with the chords, we then flipped the whole truss over the jig and put in the same number of nails on the opposite side. (It is important to verify that each joint has the required number of nails and glue, and that the holding power of the nails is not compromised due to a split in the wood.) Finally we had a completed truss.

Other Discoveries

Sometimes the wood in the pallets was so hard it required pre-drilling all the nail holes. When the strut and the slats were both hardwoods, we used a 7/64-inch (2.8-millimeter) drill bit, drilling well into the strut but not the full depth of the nails. When only the slats were hardwood, we used the 1/8-inch (3.2-millimeter) bit, drilling only through the slat and not into the strut. If you find pallets made of softwood, you can use a nail gun and compressor to speed up the process, but this should only be done with skilled operators.

Make sure that the trusses have sufficient strength for the loading conditions. A local building department or structural engineer should be consulted to find out the required design loads. If possible find an engineer with a truss-modeling program to help you with the design. At a minimum I design for 20 pounds per square foot live loads and 10 pounds per square foot dead loads when using an R-40 insulation. This is in conditions where there are no snow loads or extreme wind loads.

It is possible for you to use people to simulate the loads in order to test the strength of your truss. First calculate the amount of weight your truss will have to

CATHERINE WANEK

1-13: *Making trusses from disassembled pallets in Anapra, Mexico. Note the "jig" used to hold the pieces in place for gluing and nailing. This process enables unskilled builders to create a more consistent roof truss faster.*

carry and build one. Using the appropriate number of your friends to approximate the loading placed uniformly along the member, and with the truss set on blocks where the wall bearing would be, verify the amount of deflection experienced at the center of the member when fully loaded. This deflection should be no more than the length of the span divided by 180.

We found that we could consistently find pallets that were at least 36 inches (91.5 centimeters) square, so we designed around those sizes. At times you may find larger pallets to use, so your design can use the longer chord and diagonal lengths. Make sure the pallets are not so badly damaged that the parts are unusable due to cracking and splitting.

Many different truss configurations are possible. Possible designs include a gable truss, one with a flat bottom chord and sloping top chord, and non-symmetrical shapes. Try several layouts of the diagonal braces to find the one that places the braces directly over wall or beam supports. The spacing of the trusses will also affect their loading. If you reduce the spacing between trusses, then it increases the loading capabilities of the truss.

In areas of the world where goods are shipped on pallets, discarded or economical ones can be located for building purposes. For this reason I feel the opportunity to make structural roofs out of pallets is a very viable alternative in many developing areas of the world. We have considered the idea of facilitating economic development in some areas by empowering skilled individuals to form a small cottage industry to manufacture pallet trusses. That way the variables could be better controlled and the adequacy of the trusses assured.

Profile: Sri Laurie Baker, Architect

Ayyub Malik

Formally addressed as Sri Laurie Baker or "Bakerji" by those he works with, Laurie Baker is an English architect who has lived and worked in India for over half a century. Born in 1917 he studied architecture at the Birmingham School of Architecture and served as an ambulance driver in Burma and China in the Second World War. On his return in 1944 through Bombay — now Mumbai — he met Mahatma Gandhi. Gandhi's simple and frugal approach to life appealed to Baker's own Quaker upbringing. A year later, when the rest of his compatriots were getting ready to leave India after ruling for over a century, the young idealist Baker returned to India to work with the International Leprosy Mission to help look after their building needs.

For the next two decades he helped his wife (a local doctor) and, while looking after the medical facilities, learned local methods of construction and materials. In the beginning he felt "less knowledgeable than the village idiot, for he seemed to know what a termite, a monsoon, and black cotton soil were." Baker had brought his manuals and reference books with him, "but a bundle of comic strips would have been as helpful."[1]

What struck Baker most was the widespread poverty and deprivation, "seeing millions of people living a hand to mouth existence made me come to abhor all forms of extravagance and waste."[2] More than anything else, it is this — and perhaps the Gandhian assertion that an ideal house in the village would be built from materials obtained from within a five-mile (eight-kilometer) radius of the site — that has shaped Baker's work. In his view of architecture, "to be small is not only beautiful but also essential and even more important than large, and if we architects are even to start coping effectively with the real building problems and housing needs of the world, we must learn to build as inexpensively as possible."[3] In so doing, "… all efforts to put on a big show or indulge in deceit to make ourselves look greater than we are, seem to be quite pointless."[4]

In 1968 the Bakers moved to Trivandrum in Kerala, a progressive Communist state in the south, and the first in India to achieve full literacy and healthcare. He set up his architectural practice and has continued to live an austere but very productive life, designing buildings that relate to life and culture; and to history, climate, needs, and available resources. With an extraordinary range of work — over a thousand houses, village schools, halls, clinics, leprosy centers, chapels, civic buildings, tea houses, not to speak of film studios and a fishermen's village — Baker has seldom been short of projects. But his unorthodox approach — with its emphasis on affordability, cost reduction, and personal involvement with the workmen and materials — has won him few friends in a profession based on charging fees as a percentage of building cost. Despite this lack of acceptance he has received many national and international awards.

Baker's work is not very well known outside of India, as he has not elucidated any design theory or talked much about his architecture as such, leaving his buildings and the users to do it for him. The simplicity and economy of his buildings extend to his method of work. Instead of sitting at the drawing board or meeting contractors in his small 6-by-10-foot (2-by-3-meter) office, he prefers to develop his design and details on the building site. Using salvaged paper and the backs of carefully opened envelopes, numerous options are drawn upside down to explain the ideas to the builder and the client. The most appropriate is eventually selected to become the working drawing. Sometimes the plan of the building is chalked out on the ground, and Baker walks into various "rooms" to study their juxtaposition and views to the surrounding area.

Baker has enthusiastically promoted and advocated for various building techniques and strategies through numerous papers and pamphlets, which as yet are only available in India. In his professional work and practice, he has stuck to his beliefs with a certain sense of social purpose and moral conviction rare in the profession. The only other name that comes to mind in this context is that of the Egyptian architect Hassan Fathy (see "Elegant Solutions: The Work of Hassan Fathy").[5]

In addition to his outspokenness on topical issues such as human rights, consumerism, and nuclear weapons, Baker has been a strong advocate of social housing. His views on this subject have to some extent influenced government policy on housing and rural development. In his article *Low Cost Buildings for All*, he methodically charted out the knowledge of "our backward ancestors" about how to build in harmony with land, climate, and resources. "Vernacular architecture," for him, "almost always has apt solutions to all our problems of buildings. All that is required is to go a step further with the research … to improve on what has already been accomplished."[6] For him the proper role of modernity and meaningful development in the Indian context is to use both traditional knowledge and present technology in a way that benefits not just the few, but the population as a whole.

In its very essence Baker's approach is to seamlessly merge tradition and contemporary ideas; to renew and refresh local know-how with sensible and appropriate modern means and ideas; and to build the best and most affordable buildings at minimum expense of materials, energy, and resources. The resulting buildings are familiar and comfortable, and work with rather than against local skills, culture, and climate.

GUIDING PRINCIPLES: THE TEACHING OF LAURIE BAKER

- Only accept a reasonable proposal.
- Discourage extravagance and snobbery.
- Always study your site and see potential relating to the soil, drainage, power, fuel, etc.
- You yourself get accurate site details and in situ facts.
- Every building should be unique; no two families are alike, so why should their habitation be alike?
- Study and know local materials, cost, building techniques, and construction.
- Study the energy used in the production of materials and transport.
- Don't rob natural resources; don't use them extravagantly or unnecessarily.
- Be honest in design, materials, construction, costs, and your own mistakes.
- Avoid opulence and the use of currently fashionable gimmicks to show off.
- Get your conscience out of the deep freeze, and use it.
- Look closely at your prejudices and question them.
- Have faith in your convictions and have the courage to stick to them.
- Make "low-costery" a habit and a way of life.
- Keep your knowledge up to date.
- Don't do what is not necessary.
- Trim your staff, drawings, and equipment.
- Above all, use common sense.

Most of Indian life and work, for instance, takes place in covered and shaded open areas. By minimizing or altogether doing away with rooms, corridors, doors, and windows (so common in modern Indian buildings), Baker has designed buildings that are economical to build and maintain. In one of his houses, built for a dancer and her husband, the main accommodation of the house (living, eating, and cooking) is arranged around a courtyard with a coconut tree in the middle and a dance platform as its focus. Sleeping accommodations are upstairs. It is a fluid, flexible, and enriching arrangement of spaces for real lives to be lived in.

Baker has been responsible for all the buildings on the 9-acre (3-hectare) Centre for Development Studies in Ulloor in Trivandrum since 1967. Not based on a rigid master plan, its growth over the years has been organic and responsive to the site and changing needs and resources. Although the buildings are all different, they maintain a coherence of approach to climate, landscape, materials, and attention to detail. In most of his buildings, Baker has used perforated brick screen walls — called "*jalis*" — an important element in most Indian buildings for centuries. These are economical to build, avoid the cost of providing and maintaining expensive windows, and diffuse the bright light and glare while providing the cross ventilation so essential in this hot and humid climate. In Baker's buildings half-brick-thick jali walls often curve and undulate for structural stability while creating interesting spaces and enclosures.

Constraints of means and resources have always challenged Baker's creative thinking about design, performance, and construction. Through observation and practice, he has learned important facts such as, "… the length of wall enclosing a given area is shorter if the shape is circular, and longer if the shape is square or rectangular. This is an important factor in cost reduction. … I have found the answer to many spatial and planning problems by using the circle and the curve instead of the square or straight line — a building becomes more fun with the circle."[7] Incessantly observing, checking, improvising, and inventing

Baker produces interesting plans using triangles, hexagons, circles, and spirals — not merely to economize but also to create interesting forms and cool and pleasant environments.

Baker's contribution to Indian architecture needs to be seen and evaluated against a background of India's exploding population, rapid urbanization (with nearly 50 percent of inhabitants either homeless or living in shantytowns), environmental devastation, social and economic stress, and the government's inability to respond. In addressing these immense challenges, Baker's work is based not on any particular theory in the ever-changing architectural discourse, but is embedded in his deep concern for humanity and its essential needs. There is much to learn from how he has lived, worked, and built for those who are troubled by the everyday problems of poverty, homelessness, and deprivation. In the Indian context — or for that matter for the rest of the world — Baker has consistently demonstrated the relevance of his approach to a vast section of ignored and ill-provided-for humanity.

Baker's work has also been prophetic. For more than half a century before the present environmental concerns about sustainability, ecology, over-consumption, waste, and depleting resources, Baker has been living, preaching, and practicing what is now increasingly accepted to be the only way to make the world fairer and more sustainable.

CHAPTER TWO

Speaking the Vernacular: Mud versus Money in Africa, Asia, and the US Southwest

Jean-Louis Bourgeois

Industrial ideology transforms simplicity into poverty. It redefines an absence of industrial tools and toys as a lack of them. Planting and fanning dissatisfaction, it claims that the technological modesty of traditional people is not voluntary but a reflection of backwardness, of inferiority. This attitude is attacking traditional building worldwide.

Much discussion, many attempts, and some successes have marked the transplanting of industrial techniques from the United States and Europe to Africa and Asia. But industrial culture's most important transfer has been not of technology but of attitude. Developmentalists say that modernization teaches traditional peoples how to learn, to earn, and eventually to have. Critics dispute this trinity. [1] They say that the industrialized world's most influential lesson has been to teach the traditional world how to want.

Advocates of industrialization (claiming to be practical, hardheaded, and realistic) see traditional life as exotic, archaic, and vanishing and castigate those who cherish it as sentimentalists. The modern economy seems to say "The more you have, the happier you are" — a sentiment that equates acquisition with satisfaction. But the real message is "Have more, want more." Cajoled and goaded, the acquisitive urge outstrips accumulation, and in the end, modernity impoverishes the mind more than it enriches the hand. We become victims of an insidious psychic hunger. On every rung of the economic ladder, we feel the pressure to climb. The higher we rise toward (and into) the bourgeoisie, the greater the economic and social pressures to crave still more. And as possessions accumulate and appetites swell, "enough" remains forever beyond reach. Since individual acquisitiveness increases the gross national product, so the tune goes, "everybody benefits." Thus private greed is identified with the public good. The truth is that modernists are promoting an economic fantasy — the dream that stimulating the urge to acquire is the best way to build universal happiness.

It would be foolish to suggest that people who lack enough to eat are not poor. But the modern world has persuaded the traditional world as a whole — even people at the top of local social hierarchies — that they are materially wanting. Now, however relatively rich they are in land, livestock, or grain stores, even elites feel objectively poor.

Poverty depends less on what we have than on our attitude toward what we have. It would be ethnocentric to assume that traditional peoples have always had to restrain innate desires for a higher standard of living. Such a view supposes that necessity has forced villagers to sharply curb their wants. But having equipped them "with bourgeois impulses" out of keeping with their non-industrial tools, we judge their situation "hopeless in advance" and imagine them wracked by a despair induced, in fact, by us.[2]

In reality, to the extent that specific peoples have escaped Western values, Western pity for the villager is largely an unconscious exercise in self-flattery. In order to avoid the indignity of the word "poor," traditional societies are euphemistically called "pre-industrial," "underdeveloped," "less developed," or "developing." This ostensible civility masks arrogance: the terms redefine a present modesty of means as imminent or eventual wealth. How gracious! The terms enshrine in our very language our optimism, even conviction, that the non-consuming world is upwardly mobile and shall one day join our club. This conceptual maneuver appropriates the future. Declaring tomorrow an industrial and post-industrial cornucopia, the tactic awards validity to the present in the degree to which "now" prefigures and prepares a "soon" or "someday" full of factories, vehicles, and computers. Nonsense! The industrialized world — inflicting nuclear and toxic wastes, the greenhouse effect, and ocean dumping — can more accurately be seen as "overdeveloped" societies that need to be "de-developed."[3]

"Subsistence" is another word commonly used to describe traditional societies — as in "subsistence farming" or "subsistence economy." This term distinguishes societies without high capitalization or cash crops from those whose market economies are growing. The usage is not neutral. Employed mostly by economists keen on maximizing systems — exchange of materials, services, goods, and money — the term implies a non-cooperative, shameful withdrawal or an inability to face and join a dynamic world. Further the word suggests meagerness, a sense of barely enough, of hanging on to life with terribly fragile resources.

But a "subsistence economy" could as easily be called "local," "independent," or "autonomous."[4] True, isolated, self-sufficient villagers introduced to simple manufactured goods such as mirrors, durable knives, and kitchen utensils are usually eager to acquire them. This desire is frequently presented as evidence that, free to choose, all traditional people really prefer modern ways. But as an important geographer points out, although "demand for these simple utilitarian articles often initiates certain changes in tribal life, it does not mean a rejection of traditional culture" in favor of modern economic roles.[5]

Our judgments extend beyond words, too. Life expectancy at birth is considerably lower in traditional societies than in industrialized ones. In some regions its being as low as age 40 suggests whole populations bent and gray at 35 and few surviving 50. This is a false impression, a skewing of statistics caused by not taking into account how many infants die in regions where people follow traditional ways. Children who reach the age of 5 tend, on the whole, to live lives whose health and length approach ours. This fact surprises many Westerners who balk at the idea of living without those totems of modernity — electricity and running water.

Disturbed by television footage of famine, we tend to associate the non-Western, tropical world with hunger. The link is exaggerated. Starvation is relatively rare among the billions of people of Africa and Asia. Much famine is caused more by politics than by low agricultural yields and is suffered by refugees unable to raise food as they did at home. In the average traditional village, people eat a staple carbohydrate (usually a grain or tuber) and a sauce made of a legume (such as beans). The combination forms a complete protein. In fact most traditional peoples around the world tend to have healthier diets, eat less fat and no hormone-infested meat, and use few or no preservatives and almost no chemical fertilizers or pesticides. So who is really better or worse off?

Our economy turns people into consumers by convincing them that only industrial goods and services can satisfy their needs. In an industrial society, professionals establish and define needs and then certify themselves as uniquely qualified to fill them. For example, in the United States, it is against the law to "practice medicine" without a license. Yet in traditional societies herbalists are among those who treat diseases. Judged by US standards the herbalist is degraded into a quack or, worse, a criminal.

Similarly, building has become a profession. In the industrialized world the novice studies, gets a degree, and becomes a licensed architect. The traditional builder, on the other hand, learns by being an apprentice, usually to a member of the family. Learning to "speak the vernacular" takes place locally, without formal instruction. Architects invent ways to "upgrade buildings to make them safer" — something that usually requires the purchase of industrially produced materials. Building codes make these so-called improvements mandatory and demand that an architect supervise con-

CAROLLEE PELOS

2-1: *In Djenne, Mali a plasterer works without a trowel. Adobe walls are replastered, as needed, after one or several rainy seasons.*

struction. The license originally intended to protect the consumer against incompetence or fraud is turned into a legal weapon to coerce the consumer into employing licensees and buying industrial products. Shield turns into gun. The traditional builder is outlawed.

I am not pleading for the abolition of professions, of architectural licensing, or of building codes. But the absolute authority of licenses and codes should be relaxed to allow room for alternative ways of building — some of which have worked for millennia. In areas such as the US Southwest, where there is a vital vernacular building tradition (particularly in towns and rural areas), at least two kinds of construction should be allowed. One would conform to code. The other would not — it would be identified as "nonstandard" but would not be illegal.[6] As matters stand now, building codes deny the economically marginal the legal right to live as they choose or as they must.

Between the industrialized world (committed to high per-capita energy consumption and growth) and the vernacular world (committed to a local, ecologically healthy society), there lies an enormous gap.[7] At one extreme developmentalists work to close the gap. They seek to "elevate the backward" into the modern world. At the other extreme, traditionalists want to maintain the gap or even widen it to insulate local cultures' rich individuality against the spread of homogenizing materialism.[8]

If these unmodified positions — assimilation or isolation — appear drastic, then perhaps the responsible course is a middle one: integration. Maybe it's best to implement the faith of an anthropologist discussing tribal policy in India: "We believe that we can bring them the best things of our world without destroying the nobility and goodness of theirs."[9] The problem is that in practice this faith often fails. For example, in rural Africa and Asia, a family building or replacing a dwelling sometimes substitutes cement for mud. At first glance the process seems cheap and useful. Mud can be stabilized by adding relatively small amounts of cement or asphalt.[10] To build with, say, 5 percent cement instead of the roughly 15 percent needed in much concrete seems a major economy. But in fact the use of cement can be regressive.

Cement's appeal is that it proclaims participation today in the progressive worldwide culture of tomorrow: the cement-plastered house in the village corresponds to the glass-clad skyscraper in the city — both proclaim modernity. Yet the choice of so substantial a market item over a local one means real disruption. In an industrialized country a

CAROLLEE PELOS

2-2: *Village elders in front of the mosque of Dougouba in Mali. Sticks bristling from the structure provide permanent scaffolding for replastering.*

oor" person can buy ten bags of cement with one
y's wages; in rural Africa ten days' work buys a sin-
e bag.[11] To raise the cash someone must leave the
lage to earn a wage — must turn "industrial bach-
or."[12] Furthermore the builder in cement requires
ofessional training.[13] And, after a generation of
ing cement, many villagers lose their vernacular
ilding skills. Ultimately dependence on the mar-
t, at first voluntary, cannot be reversed.

Cement's financial, energy, water, and trans-
rtation costs are so high that its use in African and
ian desert environments amounts to conspicuous
nsumption. Indeed using stabilized mud (for village dwellings, at least) is both expen-
ve and redundant, since vernacular mud construction is quite adequate.[14] More
portantly cement is thermally inappropriate: it does not insulate, and occupants roast
day and shiver at night. Yet we continue to market cement as a superior building mate-
l that offers the best of both worlds. Free and abundant, mud is declared inadequate.
d once again self-serving technology redefines sufficiency as "underdevelopment."

Must this pattern be repeated on a global scale? Or will more leaders of vision like
ius Nyerere of Tanzania arise to speak out against "European soil": "The widespread
diction to cement… is a kind of mental paralysis. We are still thinking in terms of inter-
tional standards instead of what we can afford to do ourselves.[15]

The reality is that desert architecture needs to advance back to mud. Mud is not
me inferior material that progress will replace; rather vernacular building (because
ofoundly local) is more efficient and adaptive to local climate, culture, and ecology.
ud's benefits are psychological, too. A striking example of appropriate technology sur-
ssing industrial technology, mud reduces neocolonial dependence by promoting
ltural self-respect. But it does more than that. Mud makes gentle, lovely buildings —
pressive, ecologically sound, and humane. It is also practical and ethical. With minimal
eans it shelters against nature, without abusing her.

"Speaking the vernacular" inspires hope and resists the false, enervating myth that
gh technology shall triumph everywhere.[16] Rooted in local culture and ecology, vernac-
ar building traditions largely escape the reach of bureaucrats, professionals, and the
arket. Its beauty and ingenuity remind us that simplicity of means is not poverty of
eans. Indeed vernacular architecture both bears and makes visible a profoundly rich
man touch.

2-3: *Granaries in Affala, Niger
are built in coils, an application
of a pottery-making technique
on a larger scale.*

Case Study:
Woodless Construction: Saving Trees in the Sahel

John Norton

The Fragile Relationship between Shelter and the Environment

The Woodless Construction project is located in the band of African countries that lie to the immediate south of the Sahara (in the region known as the Sahel) and includes Mauritania, Mali, Burkina Faso, Niger, and Tchad. Local geography ranges from desert in the north to savannah in the south. Climatically the Sahel is a region of extremes — hot for most of the year, with daytime temperatures at 104 degrees Fahrenheit (40 degrees Celsius) or more. December and January nights can be cruelly cold. Rainfall is low, from 4 inches (100 millimeters) in the north to 30 or 40 inches (800 or 1000 millimeters) further south. But when rain does come, it can fall violently for an hour or two, bringing flooding and damage to buildings.

Multiple ethnic groups call the Sahel their home. Although many people live at or below subsistence level, there is an increasing monetary economy. As a result seasonal migration for work to the coastal countries further south drains young people (males in particular) from the region, making work in the villages harder for those who remain.

During years of drought in the 1970s, many communities shifted from a nomadic lifestyle — with a light footprint on the environment — to a more sedentary lifestyle. Traditionally most nomads and pastoral families lived in relatively light structures; in urban area houses, earth walls and flat timber roofs covered with earth predominated. Larger populations and a shift toward sedentary living increased the pressure on natural resources (for wood for beams, battens, etc.), outstripping the capacity of the land to support demand.[1]

People's footprint on the environment is no longer light. Where once one could walk a few yards to collect wood, now one has to travel many miles by vehicle. Where wood was once free, it now costs money (whether it is obtained legally or not). Finding good wood that will last has become ever more difficult: families find themselves replacing rotten roof beams after only two or three years because of poor quality and

DEVELOPMENT WORKSHOP

2-4: Even large buildings are designed with modest-sized rooms that can be easily roofed by domes (in this case) or vaults of earthen bricks, and constructed so that rain runs quickly and safely off the roof.

susceptibility to termite attack. (In the past, good quality wood species could last 50 or 60 years.) Many of the preferred species of tree, such as the doum palm (Hyphaene thebaica), have disappeared from parts of the region, and continued demand for wood prevents tree regeneration. And much as they might wish to, few families can afford to use non-local resources such as cement, tin sheeting, or steel.

The Woodless Construction Project

The nongovernmental organization Development Workshop (DW) has been working since 1980 to develop the use of "woodless construction" — a construction method in which all structural elements, including vault and dome roofs, are made of hand-molded, unstabilized mud blocks. Woodless construction uses no wood and requires no presses, cement, or firing to make the bricks. Instead, it uses four readily available resources:

- earth for building;
- seasonal water supplies;
- local transport, such as donkeys; and
- a large and young labor force.

The Aim

The introduction of woodless construction responds to three main objectives:

- To reduce pressure on threatened natural resources in the Sahel
- To make decent, affordable, and durable building easier for the population to achieve using local resources and
- To develop skills and income-generating opportunities in the community.

Since 1990 DW has contributed to the introduction of woodless construction in Niger, Mali, Burkina Faso, and Mauritania.

Origins

Woodless construction techniques had their origin in Egypt and Iran. In the 1970s, DW learned from and worked with Hassan Fathy and local builders skilled in the ancient techniques of vault and dome building with mud bricks (see "Elegant Solutions"). DW went on to work for more than five years in Iran, where similar vault and dome tech-

---

HOW TO INTRODUCE A TECHNOLOGY OVER MANY YEARS

- Choose suitable techniques.
- Adapt the techniques to local conditions.
- Develop a training capacity and trained builders.
- Take time to listen to local opinion and to observe local ideas.
- Convince through demonstration.
- Encourage individual initiative so people will build think for themselves.

DEVELOPMENT WORKSHOP

2-5: *Earthen vault with guiding strings. This roof is built without the need for formwork, thanks to ancient construction techniques such as the "leaning arch."*

niques were still being used to roof whole towns. (Thousands of years ago both Egypt and Iran had experienced the problems that the Sahel faces today.) A "South-South" technology transfer seemed feasible, and the first demonstration buildings using vaults and domes were built in Niger in 1980.[2]

A Question of Time and Adaptation

Although the basic idea of using vault and dome building was sound, the Sahel has its own particular conditions, cultures, and needs which require understanding and respect. For example, although arid, the Sahel experiences torrential rainstorms that are often accompanied by violent winds. Thus, unlike in Egypt where there are almost no gutters, in the Sahel special provision has to be made to quickly evacuate rainwater off the roof and protect the walls from erosion.

The people of the Sahel also lacked role models for woodless construction. In both Egypt and Iran, construction techniques are passed down through apprenticeship (often lasting many years), with learning taking place on building sites and in an environment where there are many examples to learn from. In the Sahel there were no examples to learn from, little experience of good quality masonry, and pressure to respond quickly to the growing need for an accessible building solution.

Time has therefore been a major factor in the Sahel, in circumstances requiring both speed and slowness. People needed to develop building skills quickly through training in order to provide adequate housing. And both builders and the population in general needed time to observe examples of woodless construction over several years to determine whether or not the buildings would fall down or be destroyed by rain, and to judge whether or not they liked the spaces and the appearance of different buildings.

Time has also been needed to listen to what local people have to say about their building needs, to observe existing practices and buildings, to assess problems and benefits, and to make changes that would help further woodless construction techniques in the area. Working with and training local builders has, for example, provided information about the difficulties that individuals encounter in learning particular techniques and highlighted areas where a change to the actual structure can make both the buildings safer and the building process easier. In 1996 major changes were introduced to woodless vault construction that removed the potential risk of fragility during the construction process. Over the years similar modifications have been made to the shape of domes, the profile of arches, and the bonding patterns in walls (to name but a few).

Change and improvement are ongoing, and despite the fact that woodless construction was first introduced to the Sahel over 20 years ago, there is still room for improvement with the technique, training, and the form of the buildings themselves.

Training

DW designed a series of training programs and supporting teaching tools (the trainer's handbook being a key item) to ensure that each trainee benefited from a comprehensive training package that did not sideline weaker participants or leave out certain aspects. Additions over the years have included:

- The training of trainers
- The training of novice builders
- Refresher courses
- Training for site supervisors in site management, marketing, and plans drawing
- The training of technicians, architects and engineers.

The Training of Builders

The village or small town builder is the central participant in the woodless construction process — both as actor and as agent of dissemination of the techniques. Many of these builders are illiterate, and some have only limited numeric skills. The training process takes account of these difficulties in the way the techniques are presented and can be managed even in groups where people speak different languages (although that is not an ideal situation!).

Ideally, trainees are selected in pairs in their community, since it is easier to work as a team. In some instances DW proposes that all able-bodied youth in a village have training, thereby creating a distinction between the skilled and the unskilled in a community. Since people most often build using skills available in the extended family, by extending training to as many people as possible, DW increases the chance that woodless construction will be used.

Builders new to woodless construction participate in a two-part program:

- A two- to three-week theory and practice workshop. Trainees learn all aspects of the woodless construction process (without the stigma of failure) by putting up and taking down a relatively few small training structures.
- A four- to five-week training period. Trainees work on real buildings paid for by local clients who contribute materials, unskilled labor, and finishing costs. This "partnership in training" creates additional locally owned demonstration

buildings in the neighborhood, helps build local confidence, and spreads the word in the community. It also represents a serious commitment to the technique, covering 75 percent of a building's cost.

DEVELOPMENT WORKSHOP

On average 32 trainees, under the supervision of between 6 and 8 assistant trainers and 1 head trainer, take part in each workshop. (The team structure varies according to the locality — in some cases 6 builders work as a team. All trainers started out as builders and were trained by The Woodless Construction Project, and many of the best have subsequently traveled to other African countries to share their skills.) Trainers use the Trainer's Guide, which explains the purpose and content of each session of work — what training structures need to be ready, what practical demonstration and explanation need to take place, what models have to be prepared, what practical work the trainees will do, and how to ensure that participants have understood.

Trainees receive an aide mémoire, a document that visually recaps all key points that the builder should know to build safely with woodless construction. All the handbooks use a mix of illustrations, photos, and minimal text so that literacy (while helpful) is not essential. There are some local language versions of the handbooks, but many of the spoken languages are not matched by the written word. Besides, the handbooks cannot substitute for hands-on experience and are only provided as part of the training process. Nearly all builders receive the offer of a refresher course a year or two after their initial training, particularly when there have been major changes in the techniques.

2-6: The test of successful technology transfer is when buildings are constructed without the input of the original importer of the innovation, as demonstrated by this "spontaneous" mosque.

Training is mobile. DW takes the training to the community, village, or town and runs the program in a manner that responds to local realities, uses local materials and labor, and involves the community. Woodless construction builds on local skills and experience, rather than replacing them. So, for example, we adapt woodless construction techniques to local brick sizes and use local wall-plastering techniques whenever possible. Different builders adapt the appearance of woodless construction to incorporate local and even individual styles and taste, and one can often identify the mark of an individual builder by sight.

Following more than ten years of training, local teams become responsible for day-to-day training, while DW monitors the content and quality of the work. In a country like

Niger, where there is now a long history of woodless construction, the entire program is managed locally.

Working in the Community

The measure of the success of woodless construction is the degree to which spontaneous woodless construction takes place with no external support from DW or local project staff. Thus trained builders become free agents who are actively encouraged to go out and find their own clients. They are also encouraged to build their own houses.

Some builders remain to work in their villages, which is ideal, but as many split their time between work at home and work on woodless construction projects elsewhere that generate an income for them. Such a mix of clientele contributes to the local economy and broadens the experience of the builders. It also helps move woodless construction into the mainstream of local and national building practice. As builders move out into their societies, more and more people become involved in spreading woodless construction as a viable building solution. Local families needing homes and communities wanting facilities such as grain stores, literacy centers, or kilns for making pottery (these can be built with woodless construction, too!) learn how to incorporate woodless construction in their buildings. Public sector workers (NGOs and government workers) build clinics, hospitals, and schools. Parallel projects working on rural development, food security, or environmental management add woodless construction to their programs, identifying it as a major contributor to good resource management.

2-7: Woodless Construction encourages the use of local transport, generating local income. "Local" often means the distance a donkey can carry materials.

Today there are many hundreds of woodless construction builders and many hundreds of buildings spread throughout the Sahel that were built using woodless construction techniques. In many places woodless construction has become part of the local building vocabulary. Just as importantly, pride has been restored to the Sahel, both for its builders and for the people who have taken a hand in producing quality buildings using their own skills and local resources.

Woodless construction is undoubtedly a success story in the Sahel. It is sustainable, environmentally sound, and in the hands of the local population. There is just one problem! DW and its partners need support. In a region covering half the width of Africa, there are simply not enough trained builders. To have an impact in every village, we need to train many more builders.

Case Study:
Natural Building in Thailand: From the Earth, a Village Is Born

Brendan Conley

In the mountains of northeastern Thailand, a quiet revolution is taking place. A diverse group of people who came together to construct a sustainable village has found that they are building much more. The group includes Buddhist monks, Thai students and professionals, villagers displaced by development projects, and farang (Westerners) — several hundred people in all. They spent a month in rural Chaiyaphum province, building Mun Yuen, a sustainable, self-sufficient community, and the nation's first earthen village.

"We are building a community, not just houses," said Thanai Uthaipattrakoon. He quit his job as a conventional architect to teach and learn natural building. "I want people to know that they can design and build their own home," he said. As he spoke, a group of saffron-robed monks passed by with wheelbarrows full of earth. They deposited their load in a shallow pit where young, sunburned Americans and Europeans joined Thai students in mixing clay mortar with their bare feet. Above them, a local villager balanced atop a wall of adobe as he wiggled the last earthen brick into place. "This is the way to learn about natural building," said Uthaipattrakoon, looking around. "You learn by doing, by really experiencing it. At first I thought my role would be to help design the buildings, but now I am really getting my hands dirty, working the hardest I have ever worked." Uthaipattrakoon smiled as he looked up at the first half-finished building, a community kitchen and meeting hall. "This building is beautiful," he said. "But in a way, the structure is not as important as the knowledge and spirit that we are building together."

JANELL KAPOOR

2-8: *Thai builders use mud bricks to build houses quickly.*

Bahn Din

As Thailand moves further down the path of Western development, the need for alternatives is becoming hard to ignore. Bangkok, one of the world's most polluted cities, is a sprawling, haphazard metropolis with massive daily traffic jams and few open spaces. Corruption pervades the country's military-dominated government, and IMF-sponsored mega-development projects are extracting Siam's natural resources, leaving polluted air and water behind. As the country becomes deforested, its famed biodiversity is rapidly eroding.

For Janell Kapoor, one alternative is obvious: build with mud. "I think we all have an awareness that the world we're living in doesn't make sense," said Kapoor, an Asheville, North Carolina-based natural building instructor. "Whether it's being stuck in traffic, having no time to play with your kids, or seeing violence on television, we all realize that something is not right." For Kapoor, these small imbalances are symptoms of the accelerating spread of capitalism and consumerism — the process of corporate globalization.

"The work we're doing here is part of what you might call a localization movement," said Kapoor. "Look at a conventional house in the US or Europe and try to track every part of that house — where the materials came from, how they were created, how all the machines and tools were made. By the time you're done, you'll have traveled the world, strip-mined mountains, clearcut forests, exploited workers, and polluted the Earth, all to build a house. On the other hand, look at how we're building these houses. We're using clay from right next to the site, bamboo and rice husks harvested nearby, rainwater, and hand tools. Everything is local."

At least three construction techniques are being used here to create bahn din — mud houses. Adobe bricks are made by mixing earth, rice husks, and water. Wooden forms are used to shape the mud into bricks, which are left to dry in the hot climate. A thinner mix is used for mortar. Cob construction is different: a thick mud-straw mixture is sculpted by hand in layers to form walls. The wattle and daub method is used to fill in walls between wooden posts or columns of bricks: a weave of split bamboo or branches is coated with mud plaster. Here in Mun Yuen, thatch roofs provide shelter from sun and rain.

The villagers here were displaced from their former homes and faced with building anew. They wanted to avoid contributing to deforestation, and they wanted to build simply and cheaply, to avoid adding to their debt. After Kapoor led an earthen building workshop at Wongsanit Ashram near Bangkok last year, the villagers decided that bahn din structures were the answer.

JANELL KAPOOR

2-9: *The use of cob, a sculptural earthbuilding technique that results in monolithic walls, allows for such beautiful details as these arched windows for this bahn din (mud house) at Wongsanit.*

The Struggle

For Noi Singna, one of the villagers here, the road to Mun Yuen has been long and hard. It began for her 13 years ago, with the construction of Lam Khan Choo dam. The huge government development project would destroy her home. "Our life before the dam was good," said Singna. "We supported ourselves by fishing and collecting bamboo shoots and vegetables from the forest."

Plua Chamnan, 70, also lived in the vicinity of the dam. "The government told us that when they built the dam they would also build an irrigation canal. They said we would be able to grow more rice than ever before," she said. "But this was not true. They never built the canal, and the whole area was flooded, so nothing could grow."

Faced with the destruction of their livelihood, the homeless villagers traveled to Bangkok, where they intended to press the prime minister for compensation. When he refused to meet with them, they organized a peaceful invasion, scaling the walls of the Parliament building. Riot police repressed this demonstration, knocking protesters from the walls to the ground, and beating and tear-gassing the villagers, including children and elderly people. In the aftermath, 225 protesters spent three days in jail.

Still homeless, the displaced people set up camp in front of Parliament, fasting and demonstrating for eight months. They were joined by people displaced by two other dam projects, and supporters organized by the group Assembly of the Poor. Finally, following a change in government, the protesters were offered a loan to purchase land.

The government gave us a 7 million baht (approximately US$160,000) loan to purchase this 570-rai (220-acre [89-hectare]) area of land," said Singna. "We decided to set up a cooperative to accept the loan," she added. "This gave us more legitimacy in the eyes of the government, and it made our group stronger. As individuals we had no power, but we learned through our protest that collectively we had power."

The struggle did not end with the purchase of the land. A forest fire burned the area recently, one that the villagers believe was intentionally set. Illegal logging takes place on national forest land nearby, and the loggers perceive the villagers as witnesses to their crime. When an observation tower was built at Mun Yuen for stargazing, loggers burned it down, believing that it was being used to spy on them.

With the arrival of dozens of volunteer builders and the construction of the first clay walls, the village has reached a turning point. "It's such a warm feeling having all these people here, working and exchanging knowledge," said Singna. The community is unique in Thailand. In a way, the European-American concept of an intentional community has been imported here. The village will serve as a demonstration and learning center in the future, open to students of ecology and natural building. The residents have begun a reforestation project, planting thousands of fruit and hardwood trees. Mun Yuen seems certain to live up to its name — the words mean "long-lasting."

2-10: *Working with their Thai counterparts has been an invaluable lesson in intercultural understanding for these American volunteers.*

The Spirit

Phra Sutape Chinawaro, a monk who is teaching Buddhist meditation to the builders here, knows something about struggle. As a member of the Communist Party in the 1970s, he joined a guerrilla army to fight for a Marxist revolution in Thailand. After the revolutionaries reconciled with the government, Chinawaro worked as a secular activist, and then became ordained as a monk, to work for change in a much different way.

"I discovered that if people use violent means, they will never be done with fighting," said Chinawaro. Peaceful change, he said, begins with looking inward. "If you want freedom from capitalism, you need freedom of mind," he said. "If you want a peaceful community, you must have a peaceful heart."

The idea that inner change is necessary for social change is at the heart of engaged Buddhism, a philosophy that pervades the project here. Indeed one of its foremost proponents, Sulak Sivaraksa, a Thai social critic, founded Wongsanit Ashram, which supports the building project. Sivaraksa, author of *Seeds of Peace*, has been imprisoned and exiled for his criticism of the Thai government. He promotes a philosophy of social change that is radically opposed to corporate globalization and "the religion of consumerism," and deeply rooted in the Buddhist ethic of self-awareness and mindfulness.

Buddhism is central to Thai culture, but the spirit seems to have affected the farang here too. Far from the Western missionary attitude, the foreigners are here to learn. "I'm still detaching from a very materialist, consumerist way of life," said Eliana Uretsky, of Berkeley, California. "Here, I feel like I'm learning how to be a human being."

"I'm gaining a much greater presence of mind about my role in my own community," said Julie Covington, of Asheville, North Carolina. "In the past, that was a passive role. Now I feel a need to be active, to pass on this sense of community."

For Katherine Foo, a Wellesley, Massachusetts resident now volunteering at Wongsanit Ashram, "Buddhism provides a philosophical framework for activist work" a spiritual motivation that is missing from secular organizing.

Phra Chinawaro believes that some great motivation is necessary to stop the current large-scale exploitation of people and the Earth. "The American capitalist empire is infecting the whole world," he said. The struggle of displaced people against government corruption and the building of sustainable communities are signs of hope, he said, but the journey toward peace and justice begins in each individual.

Indeed the infectious consumerism that drives corporate globalization is rooted in individual desire, multiplied by cultural and economic forces. Buddhism, with its ethic of selflessness and non-attachment, offers a way out.

Seated on the ground, Chinawaro glanced up at the adobe wall towering above him and reflected for a moment. "There are people who know the difference between the bad society and the good, and they have the ability to choose, to act," the monk said. "They have a great responsibility, and I place my hope in them."

Natural building projects in Thailand and the US are ongoing. For more information, see <www.kleiwerks.com> or <www.sulak-sivaraksa.org>.

Profile: Elegant Solutions: The Work of Hassan Fathy

Simone Swan

Visionary developers throughout the arid regions of the less-industrialized world have looked to Hassan Fathy's ideas and example, as presented in his famous book Architecture for the Poor *(University of Chicago Press, 1972), where he presaged much of the Appropriate Technology movement that now is a standard element of grassroots development philosophy around the world. Fathy's commitment to the poor, unfortunately, made him an outsider in his native Egypt, one who was regarded as a threat to vested interests in industrial building materials, banking, real estate, and large-scale contracting. Except for commissions from his friends and admirers of means, his career became almost as notable for the obstacles he encountered as it was for built work, perhaps to an extent unmatched by any other architect of his stature. —Editor,* Aramco World Magazine

After witnessing the ugliness of a peasant village and the poverty of its residents on land owned by his father, Hassan Fathy felt "…terribly responsible. Nothing had been done out of consideration for the human beings who spent their lives there; we had been content to live in ignorance of the peasant's sickening misery. I decided I must do something." Thus began his quest for a means of rebuilding communities that would allow people to live with self-respect despite their exclusion from the consumer economy. He never turned away from this goal, and the economically dispossessed were to be Fathy's constant preoccupation.

As Fathy realized that people who possess no cash can hardly become an architect's clients in the usual sense, and that they cannot be simply integrated on command into a cash economy, he set to work devising techniques of producing low-cost, energy-efficient houses. Using concrete, so much in vogue in Egypt at that time, was out of the question: It required skilled labor, expensive equipment, and industrial materials produced abroad, all of which put it well out of reach of the budget of the Egyptian peasant *(fellah)*. Worse, in hot climates, concrete traps and holds high temperatures unbearably, exactly the opposite of traditional earthen interiors, which remain cool during the day and release warmth at night.

Fathy's solution was to turn to sun-dried bricks made of mud and reinforced with straw: adobe. He engaged the advice of structural engineers and soil-mechanics specialists to determine the maximum strength and durability of adobe under different conditions. After this research, in the early 1940s, he began to design dwellings that demonstrated an unprecedented degree of harmony with the natural environment, climate and local culture, and the spiritual tradition of Islam. With inspiration from the very soil of Egypt, he aimed to help the poor build for themselves.

Yet roofing remained a problem. In rural Egypt, the *fellahin* could afford neither wood nor corrugated galvanized metal for roofs, nor could they even buy the wood needed to make forms to shape vaulted adobe roofs. Fathy's early attempts at building adobe vaulting without wooden forms — the only economically sensible solution — resulted in a series of discouraging collapses. This was particularly maddening because it was clear from his visits to Upper Egypt that just such form-less vaulting had been used for millennia to build ordinary houses, tombs, and even royal buildings.

Fathy feared that the secret had been lost. But in 1941, in the Nubian village of Abu al-Riche, he found village masons building catenary vaults of mud brick that could measure two stories high, up to 10-½ feet (3 meters) wide and of any desired length, without forms. The technique, he was exhilarated to learn, was simple enough to teach to any willing person.

Henceforth adobe become Fathy's technological passion, and he remained loyal to it not only because of its durability over millennia — some adobe structures in Egypt are more than 3,000 years old — but also because of its thermal properties. In many desert climates it maintains comfortable temperatures within a range of five to seven degrees Fahrenheit (three to four degrees Celsius) over a 24-hour cycle. Furthermore it is plentiful: approximately one-third of the world's people already live in houses made of earth. Finally the flexibility of a material for which right angles and straight lines are not always essential nourishes architectural creativity. Under Fathy's control adobe led to simple captivating beauty.

Besides using adobe to enhance thermal comfort, Fathy also experimented with the revival and modern adaptation of three time-tested vernacular architectural elements that also affect perceived temperature — the courtyard and its breezy *claustra*, or pierced wall; the *mashrabiyya*, a carved wooden window screen; and the *malqaf*, or windcatch.

In traditional desert architecture, the most efficient air conditioner available is the inner courtyard. It traps cool night air and releases it gradually during the day to adjoining rooms through built-in *claustra*, an effect that complements the thermal properties of mud brick. Trees, shrubs, and other plantings (both in the courtyard and, to the extent possible, immediately outside the house) help clean the air and afford a measure of protection from the dust-laden desert winds — or the fumes of trafficked streets. In almost all of Fathy's designs, the courtyard was a central feature.

The *mashrabiyya* is an artful lattice of lathe-turned dowels that intersect at carved

MELISSA MALOUF

2-11: *An interior view of Hassan Fathy's only US project, the Dar al Islam mosque near Abiquiu, New Mexico showing vaulted ceiling and claustra ventilation system.*

spheres (or, on occasion, other shapes). It is used as a window covering from Morocco to Pakistan. The mashrabiyya allows air to circulate through the house while maintaining privacy for its occupants, and in regions of intense sunlight it is the most effective of window shades because the curved, often polished surfaces do not block light; rather they diffuse it into the interior with the splendid subtlety of radial reflection.

The malqaf or windcatch originally developed in Persia is another millennia-old popular cooling device that fell into disuse in the Middle East when European housing design gained popularity. The malqaf is a shaft rising above a building, open to face the prevailing wind. Functioning as the opposite of a chimney, it catches and channels the wind down into the cool, lower reaches of the interior, often across a pool of water and occasionally also over wet fabrics or screens, both of which further decrease the air temperature by evaporation.

In 1985 Fathy was awarded the first Chairman's Award of the Aga Khan Award for Architecture. He was 85, and it was only then that he began to receive the international recognition, the speaking engagements, and the other awards that his work and his principles had so long deserved. On November 30, 1989 Fathy died in Cairo, in the 17th-century Mamluk house where he had lived for decades.

Patronage for Fathy's architecture for the poor never materialized to any significant degree, and his deepest hopes went largely unfulfilled in what at first seems to be a lifetime marked by setbacks. Yet Fathy remained ever an optimist, an idealist, and a fervent believer in the essential goodness and the ultimate perfectability of the human being. "Straight is the line of duty, and curved the path of beauty" are words Fathy would often mutter while drafting — words that he came to understand well in the full course of his life and work.

Today there are two centers in France inspired by Fathy. Both work with owner-builders in West Africa and the Middle East: CRATerre (Centre de Recherches en Architectures de Terre) of Grenoble and the Development Workshop of Lauzerte have helped introduce the Nubian technique of mud-brick dome and vault construction among villagers in Mali, Niger, and Iran. In Egypt Fathy's ideas can be found in the work of architects, planners, and cultural developers in numerous institutions. In the United States I have spent much of the past decade among architects, architectural conservationists, and soil engineers dedicated to continuing his work in the desert climates of the Americas. Since 1994 my resolution to carry on Fathy's work has led me to form the Swan Group in the border cities of Presidio, Texas and Ojinaga, Chihuahua.

As our global population continues to rise, the number of people without dignified, healthy, safe housing has soared far beyond what it was 30 years ago when Fathy wrote *Architecture for the Poor*. Fathy's designs, ideas, principles, and character promise to grow only more relevant with time.

FAME AND FIASCO IN NEW GOURNA

Simone Swan

New Gourna, once the jewel among Fathy's built works, is today in disrepair and largely abandoned. It is nevertheless a constant source of inspiration and an object of study for architects, planners, historians, and humanists from all over the world.

The deterioration of this extraordinarily beautiful pilot project, commissioned by the government of Egypt, began shortly after its construction. The government had neglected to take into account the psychology of what was, in effect, a forced relocation. The 7,000 villagers of Gourna had for centuries made their livings by looting the pharaonic tombs beneath the village. The decision to relocate them to a new village a few miles distant, made in the 1940s by the antiquities and housing authorities, was aimed at saving Egypt's patrimony by encouraging the villagers to become farmers.

It was quickly apparent that this created a problem that architecture, however sensitive or attractive, could not solve. The Gourna residents had expertise and knowledge of the market in tomb-robbing, but none in agriculture. They found farming less attractive and less lucrative than their previous professions. So no sooner was New Gourna built — school, mosque, marketplace, and even a theater, in addition to dwellings — than the families' breadwinners began returning to Old Gourna to continue the trade they knew. New Gourna was soon abandoned.

Today there are occasional rumors of a patron who might restore New Gourna to its original condition, but they have all dissipated without results. New Gourna stands as a magnificent object of architecture and as a rare example of conscientious planning for low-cost housing. It is referred to in classrooms throughout the world, and the political and social lessons to be drawn from its demise are no less instructive.

COURTESY SIMONE SWAN

2-12: *The adobe mosque at New Gourna, Egypt.*

CONTINUING HASSAN FATHY'S WORK IN TEXAS AND MEXICO

Simone Swan

In the spirit of Hassan Fathy, I trained my workers (The Swan Group) to build vaulted and domed roofs of adobe bricks, which are ideal for desert environments. I learned the method from Hassan Fathy himself and his masons who, in turn, had been trained not in academia but by humble dwellers of the village of Abu-al-Riche near Aswan, Egypt. My Project Manager Jesusita Jimenez of Presidio, Texas has passed on the skill to owner-builder Daniel Camacho, an adobe brick manufacturer in Ojinaga (across the Rio Grande from Presidio), and to other workers who showed an interest in this innovation. Workshop participants originating from California to Maine have learned to build earthen domes at workshops organized by The Adobe Alliance, the nonprofit educational corporation that I established in Texas.

I brought the construction techniques of woodless roofs — similar to those taught by John Norton of Development Workshop (see "Woodless Construction") — strictly for economic reasons to help local people in need of housing. The roofs are labor intensive (important, since this border region on edge of the Chihuahuan Desert is plagued by over 30 percent unemployment) and made of an inexpensive, sometimes free, materials — namely earth mixed with straw and water. These roofs in the desert yield coolness, sturdiness, and great savings. Their beauty is a pure bonus. Built without wood, they do not contribute to the devastation of the forests in the Mexican Sierra.

We have built six structures in the Big Bend region of Texas where building activities last about seven months a year during comfortable weather. From May to October the intensive desert heat tends to inhibit construction, as the mud of mortar and bricks dries too quickly and cracks. Clients have appeared from the ranks of professionals and artists; those of modest means tend to prefer purchasing trailers into which they can move quickly (but alas are frequently repossessed). On the Mexican side many women have inquired about the houses Jesusita and I built, but most cannot obtain building loans from banks, which currently lend at 48 percent interest. The Swan Group is now researching no-interest loans so to help these families have decent homes of local earth.

COURTESY SIMONE SWAN

2-13: *A detail of the adobe brick home of Simone Swan under construction. Note the arched opening and concrete bond beam.*

CHAPTER 3

Sustainable Building as Appropriate Technology

David A. Bainbridge

> *"A thing is right when it tends to preserve the integrity, stability and beauty of the biotic community. It is wrong when it does otherwise."*
> — Aldo Leopold.

Introduction

"Appropriateness" reflects the ability of a society to produce, use, repair, and dispose of technology and materials without disrupting the society and its natural environment, limiting future options, or harming future generations. But the very notion of appropriate or sustainable technology is foreign to most industrialists, designers, and engineers raised with the notion that all that is new is good. This unfailing faith in technology is increasingly eroding, however, as materials previously thought safe are found to be harmful to human health, the environment, or both. As people more deeply analyze the benefits and impacts of modern industrial technology, creation of more sustainable alternatives has gained momentum. Such appropriate technologies seek to solve problems, such as housing, food production, economic activity, etc., while doing no harm to the society or the environment. These alternatives, while rooted in tradition, make use of selected modern inputs as well, to create a hybrid approach to help solve the unprecedented problems of our age.

Through his seminal text, *Small Is Beautiful* (Blond and Briggs, 1973), economist E.F. Schumacher helped encourage a reevaluation of technology, international assistance, and development. His work ultimately led to the Intermediate Technology Development Group (ITDG) and other consulting and educational programs that began to develop technological solutions that better fit local conditions. Until these groups' arrival on the international development scene, advice to local communities had been (and, sadly, often still is) appallingly bad, developed by so-called experts who actually had a very poor idea of the challenges facing the people they intended to help, and little or no knowledge of their local resources and cultural and environmental limitations. In contrast, ITDG and other like-minded groups stressed the importance of finding locally adapted sustainable solutions to complex problems, based on a clear analysis of the situation, and doing so in partnership with the local groups themselves.

ITDG discovered that before we can develop appropriate and sustainable solutions, we need to understand the causes of problems, not just the symptoms. First and foremost one should ask the right questions: What is the problem? Is it economics? Resources? Culture? They found that asking such questions would help avoid basic errors, which until then had resulted in well-intentioned aid projects that were disastrous to local people. ITDG's and others' work led to two basic principles: do no harm; and empower people to find their own solutions. This is the challenge for builders today!

Sustainable Building Materials and Building Systems

For most of our existence on Earth, we have lived without architects, engineers, designers, or manufactured building materials. In many societies almost everyone knew how to make most of their own tools, clothes, and homes. Many of the design solutions these cultures developed were remarkable for their efficient use of materials, beauty, and longevity. They learned from observation, testing, and practice. The grass dome houses of the Great Plains Indians are a good example. Used from Kansas to Texas, this light wood frame with grass thatch was energy efficient, comfortable, and easy to build. The construction process served as a community building exercise, as well. When a house was needed, the leader would leave a twig at each house that was to provide a prepared roof rib. In the morning the "project coordinator" would establish the site, and then all the designated families would plant their rib in the designated hole. In the middle, a man climbed an oak stump with stubs for climbing and used a rope to bring all the ribs together for lashing. Then the horizontal ribs were attached and ten inches of thatch tied on. The house was finished by midday and celebrated with a feast cooked inside.

We can find hundreds of other innovative and sustainable solutions to the challenge of shelter around the world. The vernacular architectures of China, Korea, and Japan make excellent use of mud, straw, bamboo, and paper. The mud houses of Africa and elsewhere are often stunningly beautiful and tolerably comfortable. In Europe straw-clay walls are still in use after hundreds of years. The reed houses and churches of the marsh Arabs in southeastern Iraq are remarkable for their elegance and material

The ideal building would be inexpensive to build, last forever with modest maintenance, but return completely to the earth when abandoned.
— David Bainbridge

efficiency. Traditional Mayan houses with thatched roofs and thick limestone walls were comfortable, and the building shell resisted even the strongest hurricanes. Worldwide use of straw-reinforced mud is worth careful review, as is the rich history of rammed earth construction.

Traditional building techniques are not a panacea for the worldwide housing crisis, however. Because of limited understanding of engineering principles and materials properties, many traditional designs are imperfect. For example, some earthen housing designs are uncomfortable in extreme climates and may fail during earthquakes (see "Low-Cost Housing Projects Using Earth, Sand, and Bamboo"). Because of these limitations, traditional materials were often abandoned once modern materials became available. Sadly, despite new understanding and new materials, many new buildings are more uncomfortable and remain at risk for earthquakes, as poorly built concrete slab or block buildings can be as dangerous as the mud or stone homes they replace. Fortunately, traditional structures (stone or earth) that are weak in resisting tension and shear forces from earthquakes can be reinforced with locally available materials. By adding wire mesh or fabric and plaster skins, or internal and external tension and shear-resisting elements to these buildings, vernacular traditions can be improved, rather than abandoned.

Successful appropriate building solutions will often use composite materials, combining materials with different properties to get better performance. (See Table 1, page 58, for comparative properties of different materials.) Reinforced concrete is a conventional example of a composite — it combines steel rods (very strong in tension) with cement (very strong in compression). Ecocomposites combine natural materials to get improved performance — as, for example, when straw (good tensile strength) is combined with mud (good compressive strength). Examples of ecocomposites range from cob (a heavy mix of straw in mud once used in England and Europe) to *leichtlehm* (a lightweight clay-coated straw mix used in parts of Germany). Integrated system design can offer the same benefits. The Plains Indian tipi or Mongolian ger both combine a strong skin cover with a light wood frame. Putting the skin under tension strengthens the structure and improves the performance of the assembly dramatically.

SUSTAINABLE BUILDING CRITERIA

Truly sustainable building must:
- Improve quality of life
- Be comfortable and esthetically pleasing
- Improve access to homeownership for the dispossessed and poorest members of society
- Use materials that are safe to work with
- Have minimal impact on the environment
- Be easily recycled at the end of its useful life
- Support biodiversity
- Be resilient to changing environmental and social conditions
- Be locally built, maintained, fixed, and disposed of safely
- Promote community-building processes
- Be energy- and material-efficient
- Be reusable or recyclable
- Be soft, safe, fun, and healthful
- Build assets
- Be socially equitable and empowering.

Table 1: Comparative Properties of Natural Building Materials (mod=moderate)

	Strength/ Tension	Strength/ Compression	Weight	Durability
Stone	Mod-Low	High	High	Long
Mud	Low	Mod-High	High	Mod-Long
Straw	High	Low	Low	Short-Mod
Bamboo	High	Mod-High	Mod	Mod
Wood	High	Mod-High	Mod	Mod-Long
Fabrics ▲	Mod-High	Low	Low	Short-Mod
Leather	Mod-High	Low	Mod	Short-Mod
Resins	Variable	Variable	Mod	Variable
Cement	Low	High	High	Long
Lime	Low	Mod	High	Mod-Long
Metal ♦	High	Mod	Mod-High	Short-Long
Plastic, etc.	Mod-High	Mod-Low	Low-Mod	Short-Mod

▲ Cotton, silk, wool, coir (coconut fiber), jute, burlap, linen, etc.

♦ Steel, aluminum, etc.

The Basics of Climatically Adapted Design

If our designs are to be correct we must at the outset take notice of the countries and climates in which they are built. One style seems appropriate to build in Egypt, another in Spain, a different kind in Pontus, one still different in Rome, and so on with lands and countries of other characteristics. This is because one part of the Earth is directly under the sun's course; another is far away from it; while another lies midway between these two. Hence as the position of the heaven with regard to a given tract of the Earth leads naturally to different characteristics, owing to the inclination of the circle of the zodiac and the course of the sun, it is obvious that the designs for houses ought similarly to conform to the nature of the country and to the diversities of climate. — Vitruvius

Orientation

The first description of passive solar heating and cooling was written over 2,000 years ago in Greece. Solar orientation was well understood then but is largely forgotten today. According to the Greeks we are no better than the barbarians, whom Aeschylus said "lacked the knowledge of houses ... turned to face the sun ...." In Greece and Rome solar techniques were widely adapted because fuel wood and charcoal were expensive. At its peak Rome had fleets of ships scouring the known world for firewood, making them as vulnerable to supply disruption as we are today.

Solar orientation relies on changes in the sun's path over the year. In the summer the sun is high overhead at noon and rises and sets north of the east-west line (Northern Hemisphere). In the winter the sun is much lower at noon and rises and sets further south. Most traditional designers knew these seasonal changes make it possible to build a house that is naturally cool in the summer and warm in the winter. In a temperate climate in the Northern Hemisphere (with hot summers and the need for winter heating), a house should face approximately south, with several windows on the south facade, fewer on the north, and even fewer on the east and west (with some flexibility in consideration of safety, ventilation, and views). A rectangle that is longer east-west than north-south is preferred. With an overhang over the south windows to keep the high summer sun out, this house will be warmer in winter and cooler in the summer than will any other shape. A house with good orientation can save money and be much more comfortable. The best choice of specific design, however, will ultimately depend on climate, priorities of the family, and material availability.

Site Modification

A house should sit where it will have good winter sun, summer shade, summer breezes, and protection from severe winter winds. If a less than ideal site is used, it can be improved by landscaping. Big shade trees to the east and west are most important. Windbreaks can be used to block severe winds and channel cooling breezes.

Porches were a standard feature on old houses and provide many benefits for summer cooling and winter heating. With storm windows added seasonally, they can also become an effective buffer space in winter and good storage and workspace. A shaded courtyard can be very delightful in the hot summer. A two-story courtyard is easier to shade and will ventilate better, but a one-story courtyard with a full arbor, vented shade cloth, or large shade tree can be very pleasant. A large, unshaded courtyard can be very hot in the summer. A fountain, perhaps with a photovoltaic-powered pump, can add critical evaporative cooling and soothing sounds to a courtyard, patio, or even interior spaces.

Fuel Efficiency

Good passive solar design will reduce the demand for firewood or other forms of energy inputs. Almost half of the wood used in the world is used as firewood, the collection of which is often a full time task for one family member. The removal of first the trees, then the stumps, shrubs, and anything else that can be burned creates rings of devastation around urban areas. More than a billion people are harvesting firewood faster than it grows, and 100 million people are now chronically short of energy sources. As fuel shortages intensify, the price of firewood and charcoal increases, often costing as much as the food it cooks. Lack of wood makes it hard to cook food properly, to purify water, and to bathe and clean clothes — all of which can increase health risks.

As firewood consumption increases and all the trees and shrubs are removed and burned, it is harder to maintain agricultural productivity. With deforestation, environmental problems such as erosion and landslides also increase. Eventually dung and crop residues are burned, and this ensures gradually declining productivity, because nutrients are not returned to the soil.

As so many people rely on these fuels, improved stoves are a crucial component of dwelling design. Highly efficient cooker designs can significantly reduce fuel use. Simple stoves can focus the heat on the cook pot, burn wood efficiently, and be built from a wide variety of locally available materials. A simple insulating "hay box" can also be used to conserve the heat of cooking for dishes such as rice or beans that can continue to cook at a low temperature. Internal flue pots can reduce fuel use 10 percent and are traditional designs in very fuel-restricted areas. And, in some cases, solar cookers would work.

A program to fund reforestation of millions of acres and to improve fuel-burning efficiency would be an excellent antidote to this trend and could be funded by "carbon credits." More efficient stoves would greatly improve indoor air quality and lessen respiratory illness caused by interior open fires. The equally important issues of land tenure, woodlot development, and protection of sensitive areas should also be addressed to reduce the world threat of the firewood crisis — the "quiet" energy crisis.

SUSAN KLINKER

3-1: *Mexican women and children engaged in producing straw-clay blocks, a sustainable building material using local resources.*

Insulation

Although orientation is the first step in any proper design, insulation is also necessary to save the heat from a sunny winter day for the cold night. Stone or mud walls are poor insulators and hamper traditional solar designs. In comparison the walls of straw-clay and, even better, straw bale buildings provide excellent insulation and allow for exceptional comfort and performance even in extreme environments. Fuel use for heating in Ulaan Bator, Mongolia, for example, was cut 80 percent in straw bale buildings compared to conventional structures.

Although insulation is usually an inexpensive component in new construction, it can be expensive to retrofit. Insulating attics or roofs is particularly important. In non-industrial areas, insulated windows can be challenging, but you can make serviceable double-pane windows by doubling up inexpensive single-pane windows. A ½ - inch (1 centimeter) gap between these windows is desirable, as a gap of over ½ an inch (1 centimeter) increases convection currents and is less efficient. Insulated shutters and drapes are very effective, while seasonally installed storm windows can also improve window performance. Skylights are energy costly and should generally be minimized to the size needed for indoor lighting, if used at all.

Weatherization

Heat is also lost or gained by airflow and may account for half of the heat loss in a well-insulated but leaky house. The goal in a well-tempered house is controlled airflow — when and where it is wanted. Super-insulation and a snug house make for superb comfort, and in most areas, this type of house works so well that open windows can be used to provide fresh air all year. In hot climates windows can be opened at night to let in cool night air, which can be held by the insulating walls throughout the day. In any well-insulated and weatherized house, air quality must be considered by ensuring that adequate air changes will take place. This is particularly important if any interior air pollution is being generated by such activities as cooking with gas or wood. Air-to-air heat exchangers can be used to reclaim the heat of exhaust air in very cold climates.

Thermal Mass

Heat from sunny winter days can be saved for nighttime, using thermal mass inside a well-insulated shell. A thick plaster layer on straw bale walls, for example, provides considerable mass, which can be augmented with an exposed concrete, adobe, or other high-mass floor, adobe *bancos*, masonry, or water tanks. Doubled plasterboard can help retain heat in frame houses, while in traditional homes stone wall facings or partitions increase working thermal mass. Thermal mass also helps save the coolness of a summer night for the following hot day. The more thermal mass a house has, the more stable the temperature will be. But without exterior insulation, thermal mass can create problems of overheating in summer and chilling in winter.

DAVID SHAW

3-2: *This high-mass masonry stove at the Earth Sweet Home Institute is an example of a highly efficient heating technology designed to minimize the use of wood and reduce pollution.*

Solar Hot Water

There are three general types of solar water heaters. Integral or passive solar water heaters can be as simple as a black can in an insulated box with a window. They are reliable and inexpensive, though if the volume is too high in comparison to the surface area, they can be inefficient. Thermosiphon systems use the reduced density of hot water to drive circulation from a flat plate collector to an insulated storage tank set above the collector (or with special valves below it). Active solar systems use a pump and controller to transfer hot water from a collector to a remote well-insulated storage tank. Active solar has the highest performance, but also the highest cost and lowest reliability.

Shading

Orientation and shading are the keys to keeping a house cool in the summer. A properly sized overhang and/or arbor will keep out hot summer sun but will let in desirable winter sun. Windows on the east and west are more difficult to shade properly. Arbors, vertical fins, shutters, shades, or trees can be used to shade these windows. If available, shadescreen is a step in the right direction and is good for retrofitting. A large window in the wrong place can make a home uncomfortable without air conditioning, and even with air conditioning the cost of operating the home will be increased — forever.

Ventilation

To capture cooling breezes in the summer requires careful window placement and interior design. Privacy *and* good airflow are possible. In many situations insulated screened vents can be more economical than operable windows. Interior airflow can be improved if doors are cut one or two inches above the floor level and vents and windows are placed above doors. In hot climates raised floors and high ceilings increase ventilation and improve comfort. Cool air for ventilation can be drawn from shaded areas near the ground and landscaping, which tend to stay cooler in hot climates.

3-3: *Wind catchers in Tatta, Pakistan. Reaching up from the roof, wind catchers capture cooling breezes and channel them down into the house.*

CAROLLEE PELOS

Other Key Issues for Building Designers

Many human needs are affected by design choices and material selections in building. Water collection, storage, and treatment (through water harvesting, cisterns, water catchments, sand filters, and solar disinfection or distillation) are some of the most important. Wastewater from washing (greywater) and rooftop water catchment stored in cisterns can provide water essential for irrigation in arid lands. High efficiency irrigation such as clay pot, deep pipe, and wick irrigation can get the most out of this available water. Proper design of water harvesting and landscaping can provide irrigation critical to help a household meet its needs for food, fodder, homebuilding materials, tools, and medicine.

Food storage is another important consideration. Inexpensive solar dryers can make it easier to store food safely. Ventilated cabinets, evaporative coolers, well houses, or underground chambers to store winter ice can also increase food availability.

Waste treatment can be provided with a waterless composting toilet or pit toilet. Greywater and wastes can also be treated in a biological system of aquatic plants and microorganisms, though aquatic plants are not necessary in all cases. These can be done at the home scale, but the neighborhood scale may be more appropriate.

Making Sure Things Work

It takes hard work, persistence, and careful consideration of the social and environmental setting to create successful solutions when dealing with appropriate technology transfer projects. Indeed more projects founder on social concerns than on technical problems. It takes time to understand and work with local communities. Most Peace Corps workers will admit that two years of residence were a great learning experience, but only at the end of the two years had they gotten to the point where they could begin to be effective.

Humility and a willingness to listen are crucial tools for development workers. Local people often have the best understanding of the causes of a problem and potential solutions, yet because they are not technically trained or accustomed to presenting information to experts or to other audiences they are often ignored or scorned. It usually takes women, as well as men,

WHY PROJECTS FAIL

The most common reasons for failure in building and technological assistance include:

- Asking the wrong questions. (You must isolate and define the problem carefully and identify causes rather than symptoms.)
- Not considering risk
- Misconstruing tenure and economic relations
- Talking to the wrong people (men only, for example)
- Not researching historical solutions
- Failing to provide maintenance support (training and supplies)
- Promoting untested solutions (prototype and test carefully; demonstrate successful examples)
- Failing to match solutions to environment and/or culture
- Neglecting education and training
- Ignoring infrastructure and marketing limitations
- Hurrying (technology transfer takes time and patience)
- Making big leaps rather than little steps (sometimes little steps are best)
- Neglecting details (the devil truly is in the details)
- Neglecting maintenance costs, and technical and material demands for repairs
- Using nonrenewable, nonsustainable energy or material resources
- Ignoring equity or fairness.

to find out the full scope of a problem and find solutions. Women agents are particularly important in cultures where it is inappropriate for women to speak to a strange man (or any non-family male).

Although progress may be slow, it can be speeded up if the causes of inertia are recognized and addressed. Inertia can result from the fact that:

- Information is only a small part of the decision-making process.
- Information is a political resource.
- Human information processing is based on simplification and biased toward experiential learning.
- Many communities reward stability rather than change.
- People and communities are understandably risk averse.
- Institutions are very effective at self-preservation.
- Even if they are very poor, people are usually busy, without free time for new "cheap, elegantly simple but labor-intensive practices."

Obstacles can be minimized through skillful planning, marketing, and management. An incremental approach minimizes most of these obstacles and works well with an ongoing adaptive review and evaluation process. Taking many small steps allows people to learn more about environmental and social systems as well as sustainable systems. Encouraging participation in this step-by-step process of muddling through can also minimize the problems caused by political mismanagement of information.

Marketing viable solutions can be helpful. Good solutions are often ignored while well-marketed, poor solutions are adopted or funded. Making information readily understandable for a particular culture is an essential part of the marketing process. You can't stop a successful technology by not marketing it — if it works it will spread — but you can introduce it much more quickly with marketing support.

Sustainable building is appropriate technology. It is something we, as a species, have done for most of our time on Earth. We need to rediscover and refine the locally adapted designs and materials that make housing efficient, comfortable, and sustainable. Sharing the best ideas from around the world and improving them with the current understanding and insight of science and engineering can create new solutions that are even better then the best traditional designs. To sustain life on Earth, we need to find how to make building more sustainable. In developed countries building represents 30 percent or more of the energy and material flows; in less developed countries it may be even higher.

The goal is to increase sustainability and to improve the quality of life. Making buildings more sustainable (less costly, more efficient) can free up more time, money, and energy for education, community building, and environmental restoration. Every house makes a difference.

A Seven-Step Program for Addressing Problems with Housing, Resources, Health, Sustainable Food Production, or Environmental Restoration

1. Define the Problem

The first step is not to just describe the symptoms but to identify the causes. This is the most important step and often the most challenging. One must ask the following questions: What is the history of the problem and current status, conditions, and trends? What data are available? What are the environmental and human health consequences (earthquake risk, respiratory illness, allergies) of current practices? Usually the most important factors are related to money. What are the flows of money and resources in and out of the family and community? What subsidies exist for unsustainable materials and energy? Develop a budget of the true cost of resource use, including the value of nature's services such as oxygen generation, flood control, and pollution cleanup. Is the value of natural capital declining or appreciating? What are the tenure (access to, or rights to resources) issues? Does planting trees jeopardize ownership or use? Is fuel wood legally available or poached?

3-4: *Simple machinery, such as this baler in China, can be used to create building materials from local resources. The resulting straw bales are inexpensive, relatively easy to build with, and highly insulating.*

2. Assemble a Research Team

Identify the stakeholders and include a broad selection of them on the problem-solving team. Ask: Who is directly or indirectly involved or crucial to making changes? Who is interested? Use a wide net to capture people in interlinked systems: male, female, rich, poor, young, and old. Make the research team multidisciplinary if possible, with workers in the field, normal people, professionals, families, singles, and those with a wide range of experience. Women usually play a vital role in resource management but were once ignored in traditional aid and extension work. A couple with children is the ideal field team for understanding and solving problems. If only one person can go in from outside, a woman is often best because she can talk to the women and can often make it into the men's world as an "other" if she is accomplished in a traditional male field, such as building.

3. Research

It is important to understand the ecological setting, including climate, geology, hydrology, and natural and managed ecosystems. Collect both hard and anecdotal evidence. Seek out elders and ask: What was it like 10 and 100 years ago?

In addition to assessing the local material resources, we also need to assess local human resources and capability. This is radically opposed to the usual needs-based assessment. If we start looking at needs we can always find more needs, enough to paralyze a community or an advisor. But if we focus on resources, particularly the skills, aptitude, and attitude of local people, we can find we are very rich indeed. Toward this end try to use analyses that will help you understand the cultural history, psychology, and education of the area. Focus on the skills in the affected area, not just the limitations. Ask: What is the labor situation? Do people work away from home part of the year? How is the area administered? What is the current community educational system like? Is there an understanding and support for sustainability? Are there any charismatic progressive leaders? What is the level of literacy? Is there access to information and exposure to TV, radio, newspapers? How important are religion and religious institutions? What is the political setting? Is there a formal legal environment or are decisions made by custom or elders?

An analysis of the economic and political situation should also be undertaken. Ask: How competitive are the industries and activities in the area? How are ecosystems managed? What are the problems and potentials for agriculture, forestry, ranching, and manufacturing? What waste materials are available? How does the economy currently function? Who benefits? Who pays (local, national, global: present and future)? What is the flow of money and resources in and out of the area? What resource and tenure conflicts or traditions exist? Consider both micro- and macro-economic trends, equity, taxes, savings, investments, and subsidies. Were traditional practices sustainable? Are today's?

4. Work with Stakeholders to Develop Ideas and Draft Solutions

A participatory process (Rapid Rural Assessment and Participatory Rural Appraisal, for example) will most effectively bring problems and solutions to the fore. Ask: What is holding back problem solvers? Can problem solvers be aided or rewarded? Can problem creators be fined or charged; can victims be paid? What have other communities, cities, counties, or countries done? What worked? As solutions are developed, use a feedback mechanism to make sure that the real problem is being addressed. Always strive to first do no harm.

5. Prototype, Test, Demonstrate, and Educate

Try a small-scale project before recommending another new solution. Demonstrate a building with the advisor and his family or with a local family (preferably with children) living in it, if possible. Depending on cultural or political situations, a community building may be a more appropriate first step. Set up accurate accounting systems for money and environmental services, comfort, and health. Find out what works and what doesn't. Follow up on a project for more than a year if possible. Does it fall down? Does it become less effective? Does it work? Demonstration is critical and offers a community the chance to see if the new or, more often, revamped traditional solutions will work. As Marshall McLuhan said, "Knowledge doesn't change behavior; experience does."

When transferring technology user-to-user information transfer is best. One of the common failings of our time is the reliance on experts, consultants, and the government to fix things. They can't. Squatters in many cities can plan, take over a plot of unused land, and build a new community over a weekend. The official process can take decades.

Information is always lost in the information chain from farmer or builder to researcher to extension agent to user. "Builder to builder" and "farmer to farmer" programs are much more effective. The straw bale building movement offers many lessons about how to do this: hear one, see one, do one, teach one. Never be intimidated by the magnitude of the task — a single person with a good idea and courage can change the world. Successful sustainable building technology will sell itself, if it works.

6. Refine

Monitor and repeat the process as you gradually refine and improve solutions, publicize successes, and admit failures. Train local workers and leaders during prototyping and demonstration. Look for unintended consequences. Work toward reduced costs and improved effectiveness.

7. Report

Let others know what worked and what didn't on the Internet, through comics, newsletters, articles, plays, street theater, farmer-to-farmer and builder-to-builder exchanges, open houses, and videos.

Case Study:
New Beginnings: The American Indian Housing Initiative

David Riley

Housing shortages on reservations are severe. Those tribal members who do have homes live in overcrowded and underinsulated rental units, resulting in a myriad of health problems and excessive heating expenses, and little hope for economic stability or growth. When representatives from The American Indian Housing Initiative (AIHI), presented demonstration straw bale homes to reservation communities, tribal members were quick to recognize the advantages and potential of straw as an alternative to conventional construction.

Led by faculty at Pennsylvania State University and the University of Washington and their partners at the Red Feather Development Group, AIHI engages tribal members and nonprofit organizations in community-based building projects. Because of the volunteer friendly construction process and the abundant supply of straw on the northern Plains, straw bale construction allows tribal members to build themselves comfortable, durable, energy-efficient housing. As of the summer of 2002, AIHI has constructed three load-bearing straw bale demonstration homes and two community buildings on various reservations. The ultimate goal of the initiative is to help demystify load-bearing straw bale construction in the region and to build the capacity for tribal members to build structures of their own.

The Process

AIHI projects have all been accomplished through summer "blitz builds," in which an advance team prepares a foundation and deck, and a larger group of students and volunteers constructs the building in the span of two to three weeks. Courses at Penn State and the University of Washington prepare students for the summer construction experience. Tribal members and other volunteers join the students on-site. The annual cycle of preparation, doing, and reflection allows students in architecture and engineering disciplines to develop and refine model home concepts, straw wall-raising techniques, and design details to be used in future projects. Each year lessons from previous projects are incorporated into the design and building processes.

Model Home Design

Design details and methods appropriated for these projects remain well within documented performance of laboratory testing and existing straw bale building codes. Three

goals guide the design and subsequent refinements and improvements to the design:

- Demonstrate the viability of load-bearing
  straw bale building methods as a regionally
  appropriate sustainable housing solution.
- Maximize the inclusion of locally available sustainable building materials and
  the use of semi-skilled labor (trained on-
  site) through teachable design details and
  building methods.
- Minimize expensive building materials,
  technical construction steps, and the
  connectivity between community-built
  elements and parts of the building that
  need to be completed by contractors.

MICHAEL ROSENBERG

3-5: *Prefabricated modular plywood frames are used to create closets, shelves, and door openings in straw bale walls.*

Three demonstration homes on the Crow, Lakota, and Northern Cheyenne reservations were used to experiment with load-bearing straw bale wall systems, and a
whole-house design concept that would be both sensitive to the values of tribal members, while also appropriate for the harsh climate of the Northern Plains.

The first project, a 1-½-story, three-bedroom home was constructed in May 1999 on
the Crow Indian Reservation in Garryowen, Montana. The project represented the first of
its kind in the region, and the first opportunity for students to identify what materials were
specific to the Northern Plains. It was quickly determined that only smaller two-string bales
(which have limited load-bearing capacity) were available in the region. To allow for a
wider building, a double-bale (two-wythe) wall system was devised with the help of Chris
Stafford, a straw bale architect and builder. The 3-foot-wide (1-meter-wide) wall permitted
us to use longer spanning trusses and to devise a more functional floor plan.

Horizontal strapping of bales kept walls stable and aligned under vertical loading.
The strapping was then passed through the walls to help tie the two wythes together.
Students created construction drawings and devised a daily construction schedule for
the two-week blitz build. At the conclusion of the project, which required much more
time to complete than expected, students assessed the project and subsequently produced alternative plans based on their evaluations. These plans included a simplified,
single-story floor plan and workshop methods to make the construction process more
inclusive of the community.

The second project, a single-story two-bedroom home, was constructed in the summer of 2000 on the Pine Ridge Indian Reservation in Red Shirt, South Dakota. The improved design used the concept of a minimalist utility core surrounded by an easy-to-construct straw bale shell.

To make the wall-raising process more systematic, students and faculty developed a more detailed set of construction drawings and assignments for wall captains. Construction schedules and illustrative workshop handout materials were also created for the project, with the specific goal of achieving a 14-day completion time.

The improvements to the design did decrease cost while increasing the percentage of community and volunteer labor and decreasing the technical steps in the construction process. The core and shell of this project were successfully completed in 14 days.

At the conclusion of the project, students proposed further changes and improvements to the design. These included increasing the size of the home to match the average needs of families on reservations, and generating additional drawings and specifications to describe the design to the various groups that would take part in a community-built project.

A third demonstration home was constructed in July 2001 for the Bear Quiver family on the Northern Cheyenne Reservation in Busby, Montana. Inspired by the straw bale home on the nearby Crow reservation, Martha Bear Quiver secured a USDA rural development mortgage and decided to construct her home out of straw. This project was a single-story, four-bedroom house designed by AIHI partners with the help of the Bear Quiver family.

MICHAEL ROSENBERG

3-6: *The empowerment of volunteers to contribute to the building process is the greatest attribute of straw bale construction. A crowd of friends and strangers come together as one team around a common goal.*

Further refinements to the design details and floor plan were incorporated in this project. The extensive use of wood-framed interior walls and the inclusion of two bathrooms created significant problems for the volunteer labor force. For the next prototype home AIHI members are investigating the use of a prefabricated modular core that would allow a community-built shell structure to be built around a factory-finished kitchen, bath, and laundry core.

Engaging the Community

The first three projects proved valuable in demonstrating the use of straw and in establishing the presence of a straw bale model home in the community. It became clear, however, that to truly engage tribes in the process, the construction of a single home for a single family does not adequately make a connection with the tribal community. Consequently leaders of AIHI decided to construct community buildings that would benefit the tribes as a whole. Leaders also decided to provide more time for students and faculty to visit with tribal members to increase understanding of the challenges indigenous Americans face and the values they place on family, community, politics, and culture.

In the summer of 2002 the initiative partners divided into two teams and took on the construction of a 1,000-square-foot (93-square-meter) study hall on the Crow Indian Reservation in Montana and a 1,500-square-foot (140-square-meter) literacy center on the Northern Cheyenne Reservation. Both projects offered the opportunity to construct a straw bale building in the heart of the respective communities, thereby benefiting each community as a whole.

The Crow Study Hall project was led by Red Feather Development, with design and project management services provided by a graduate of the University of Washington (a veteran of the first straw bale design-build course). Base funding for the project came from four Crow teens who won a national science fair contest with their straw bale entry and decided to use their prize money to build a central location for their peers to do school work (see "Crow Girls' Winning Science Project Is *Not* the Last Straw"). Additional fundraising by Red Feather and the Universities allowed the project to be completed and operational as of August 2002. The building now serves as a quiet place to study, with new computers available for homework assignments and school projects.

The Literacy Center project, led by Penn State and the University of Washington, was constructed on the campus of the Chief Dull Knife College on the Northern Cheyenne Reservation in Lame Deer, Montana. Funding for this project was obtained through a USDA Tribal College grant, with additional resources provided by the AIHI partners. Local Northern Cheyenne artists provided artwork for the building in the form of tile mosaics and trim paintings. The vaulted ceiling and round interior walls of the building offer cozy nooks where students can read and study. As childcare is hard to come by on the reservation, a children's area provides informal daycare for students with young family members, reducing one of the previous barriers to parents taking advantage of the services provided at the center.

3-7: *Each volunteer leaves their mark in a hand-crafted home. Like clay in the hands of a sculptor, stucco walls become an open medium for expression.*

During the Literacy Center project, research teams of faculty and students visited community organizations and tribal members in an effort to assess the values and concerns of the tribe with respect to housing conditions and their communities. Evening presentations from the tribe helped enlighten the visitors to the ways of the Northern Cheyenne and provided a chance for the two groups to come together informally, building friendships and trust.

During preparations for the projects, several community groups came forward with ideas for new facilities and showed an interest in building with straw. One group had just succeeded in obtaining a USDA grant to build a new community center with volunteer labor. The project provided an excellent chance to test the use of straw bale on a project initiated and mostly built by tribal members instead of outside volunteers. Minimal assistance got the project off the ground, and with technical support on the straw bale details, members of the Muddy District were well on their way to constructing a valuable new facility for their community. The new center will house offices for tribal council representatives, a kitchen, rest rooms, saunas, a therapy room, and a large meeting room — a substantial improvement over the old center, which consisted of an uninsulated steel building and an outhouse. The center now serves as an example to the rest of the tribe of how to make positive steps to improve their communities.

The Future

Our work encourages an understanding of what a community can achieve through a shared vision and shared effort. Housing needs are intertwined with all facets of economic and community development, and there is much more to solving the housing problems of American Indians than just the construction of new homes. With this in mind the partners of AIHI will continue to build partnerships with American Indian tribes. In time tribes will have the capacity to rebuild their communities on their own terms.

The simplicity of straw bale construction allows the techniques to be passed on easily. It is the hope of AIHI partners that the exponential transmission of skills from a few leaders to entire crews will be paralleled by the exponential growth of community-built structures on American Indian reservations.

Profile:
Crow Girls' Winning Science Project is *Not* the Last Straw

Michelle Nijhuis

On a blistering morning in late July, four young girls sit on a stage in the town park, smiling and sweltering in heavy dresses trimmed with elk teeth.

In English and the Crow language, one speaker after another calls them symbols of hope for the Crow tribe. "We have just met the next generation of tribal leaders," says Crow chairman Clifford Birdinground, Jr.

It's a lot of fanfare for a middle school science project. But what started as a simple series of experiments has turned into something extraordinary. The project showed that straw bales are a safe, energy-efficient building material — and it won enough prize money to fund a straw bale study hall on the reservation.

In the summer of 2002, a crew of volunteers raised the building, and its metal roof now gleams near the edge of the park, inspiring renewed interest in straw bale building as a solution to the chronic housing shortages here.

For the 11,000 members of the Crow tribe, such tangible accomplishments are rare. The tribal government is shadowed by chaos and scandal; unemployment on the southeastern Montana reservation ranges from 60 to 80 percent; and the county is consistently among the 20 poorest in the nation.

The girls — Lucretia Birdinground, Kimberly Deputee, Omney Sees the Ground, and Brenett Stewart — "were getting a lot of flak from their peers at first," their teacher, Jack Joyce, tells the crowd in the park. "The boys in their class said, 'Why are you working so hard? You're not gonna win anything.' Well, you might say those boys are now eating crow."

In 2000, when the girls entered the eighth grade at Pretty Eagle Catholic School, Mr. Joyce told them about the Bayer/National Science Foundation competition for middle schoolers. Each entry had to use science to address a problem in the community. The girls chose one they were all too familiar with: the lack of quality housing.

On the Crow reservation, as on many others, multiple families often crowd into trailers or cheaply built government housing. The drafty houses are expensive to heat, and several people have frozen to death in recent years during the frigid Great Plains winters. Joyce suggested that the girls visit a local resident who lived in a house built with straw bales. "When he told us about straw bale houses, we kind of looked at him like he was crazy," Lucretia remembers. "It wasn't until we saw that house, and saw how pretty and nice it was, that we knew he was telling the truth."

When the girls talked to other tribe members about the advantages of straw bale houses, they met a lot of doubts and questions. So they designed some experiments to show that the stucco-covered walls are waterproof, fire-resistant, and a good form of insulation. When their entry made it to the competition's Top 10, the girls and Joyce were flown to Disney World for the final results.

They captured the grand prize: a $25,000 Christopher Columbus Fellowship Foundation grant to help them construct a straw bale building. Oprah Winfrey then flew the girls to Chicago to appear on her show. She gave them another $25,000 grant, plus $20,000 worth of Stanley tools.

The next step was to talk to the Seattle-based Red Feather Development Group, which had built the first straw bale house on the Crow reservation. The nonprofit group uses volunteer labor, and its staff was eager to begin another project in Crow country.

The girls decided on a community study hall as a project that would not only demonstrate the advantages of straw bale building but also benefit tribal members directly.

An architect from Red Feather designed the structure, and 35 volunteer builders arrived on the reservation in mid-July. They worked in temperatures near 120 degrees Fahrenheit (49 degrees Celsius) to finish the 900-square-foot (85-square-meter) building in just 2-½ weeks.

The girls worked at the site nearly every day, but many of their neighbors were hesitant. "It's a great project. I just wish there were more of our people out there," says tribal legislator Jared Stewart. "People are so afraid to try new things."

That reluctance may be fading. Dozens of tribe members toured the new study hall at its official opening, and more than a few asked how they could secure a loan for a straw bale house of their own. The Red Feather Development Group has made a three-year commitment to housing projects on the Crow reservation, and Executive Director Robert Young is talking with the tribe about establishing an independent development office in Crow Agency.

SchoolKiT.com donated computers and software to the study hall, and Rena Frank, the tribe's director of human services, hopes everyone "from three-year-olds to senior citizens" will find a use for the space. Ideas include individual tutoring, traditional storytelling, and recording oral histories.

The girls are finished with construction work for now, but they may begin other projects soon.

Lucretia dreams of building a straw bale house for her family, and Kimberly plans to become a professional builder. "I can't believe this [study hall] is real," Kimberly says. "For so long we just had it written down on paper. We had no idea it was going to come to life."

ON CREATING AN AFFORDABLE STRAW BALE HOUSE: A CAUTIONARY TALE

David Bainbridge

As one of its earliest proponents, I was attracted to straw bale building because it had great potential for providing comfortable, energy-efficient, durable, and affordable homes, schools, and buildings. Unfortunately in our efforts to meet building codes, placate nervous engineers, and avoid lawsuits, the complexity and costs of alternative materials have risen to equal and sometimes exceed conventional building materials. I therefore propose a long overdue return to common sense.

For affordable housing we need to get back to basics — we housed ourselves for several hundred thousand years without expensive and costly homes and brutal financing costs. By doing so again we can reduce the enormous environmental cost of buildings, estimated in industrialized countries at between 25 and 30 percent of the total nonrenewable material flux. It's time to step back and see what makes sense and is affordable. Somewhere between the homeless person living in a home made of cardboard boxes and the $200-a-square-foot custom home, there is an affordable home that is comfortable, healthy, energy- and resource-efficient, and affordable.

If I were in the position to build a new house, I would begin with careful ground shaping to create a dry island with outsloping perimeter and excellent drainage. I would think ahead to the 100-year flood. (If you need to build a mound like the Dutch used to, do it.) This is a critical step — much more important than trying to keep the house dry by using membranes and drains.

For the foundation, in most conditions, I would use a rubble trench as shallow as prudence allows, perhaps only 12 inches (30 centimeters) deep, with a filter fabric-wrapped drain near the base draining to daylight, if possible, or to a large dry well. Crushed concrete (also known as "urbanite") is a good rubble fill. I would compact this thoroughly. What's good enough for freeways and airports works well for buildings.

Then the choices get more difficult. If I were in a developed country with engineers or inspectors looking over my shoulder, I would add a 5½ by-24-inch-wide

(14-by-60-centimeter-wide) grade beam with a minimum of three #4 rebars. In some places bamboo or barbed wire could be substituted for the rebar. It might be desirable to add a few reinforced piers tied-in to the beam (poured in postholes) to prevent sliding around in an earthquake or high wind. In cold climates the frost-protected foundation systems from Europe could be used to prevent frost heave.

In many areas the rubble trench and/or the grade beam could be skipped, and a waterproof membrane could simply be laid on the ground. For Alliant University's seed building in the Mojave Desert we used carpet runner — a heavy vinyl readily available in appropriate widths. A well-compacted rammed earth footing could also be used, with asphalt or cement for water resistance if needed. Or dry stone or adobe — whatever is available. On a sloping site I would be tempted to use a pole foundation. This would minimize site disturbance, help ensure that there would be no wet walls on the upslope side even in heavy rain, and wouldn't require massive amounts of concrete.

CATHERINE WANEK

3-8: *An affordable straw bale house prototype, under construction in New Mexico. This house features a low-tech, recycled concrete and earthbag foundation, straw bale walls with earth plasters, and a simple metal roof supported by pallet trusses.*

The straw bale wall would then be laid on a rot-resistant or treated-wood ladder frame filled with small gravel, with drains. A pressure-treated plywood sheet on the bottom of the ladder would also be a possible foundation solution. These would be load-bearing straw bale walls, unless design choices (windows and doors, roof weight, snow loads) or regulations make a post and beam more appropriate. If columns were needed, a simple reinforced concrete column (widely used in the developing world) would work well. These could be poured in place with rebar or wire reinforcing, or built in forms on the ground.

The bottom bale would be 6 to 9 inches (15 to 23 centimeters) above grade. The walls could be tied down in many ways. A conventional approach would use rebar extending through the grade beam, or slab tied into rebar running up the walls (either in the center or paired inside and outside) at 24- or 36-inch (60- or 90-centimeter) spacing. This would then be tied into rebar extending from a reinforced 5½ inch (14-centimeter) concrete bond beam on top of the wall. Hooks or ties set in the wall top beam would be used to provide uplift resistance with the preferred wide overhangs.

This would be costly and a bit of overkill, although building codes may require it. Much simpler alternatives might suffice. If a bond beam were used, rebar hooks or strong wire ties could be left exposed to serve as tiedowns for the wall and roof. The tiedowns could be wire, polypropylene or metal straps, plastic baling wire or hemp rope or twine. Heavy wire could also be tied around large chunks of rubble deep in the footing and tied into the wall or roof. (In one building without a foundation, we used screw-in anchor disks like those used to brace power poles.)

Hand-applied lime or earth plasters would work for the wall finish. These sometimes benefit from reinforcing mesh, especially around doors and windows or places where bales meet timber. A faster but less structurally advantageous alternative would be to use painted canvas or burlap with plaster.

Roofing might be thatch (lifetime up to 75 years for reeds) or metal. Both are lightweight and require minimal framing. A metal roof would also be ideal for rainwater collection.

The interior floor (ideally a couple of inches below the top of the grade beam so interior floods won't reach the bales) could be puddled adobe, brick, tile, or concrete wall toppers over compacted sand; or plaster directly on highly compacted soil. (Thin-plastered floors from the Middle East have been found that are over 5,000 years old.) Linoleum over a dry, well-compacted dirt floor would also be possible.

Let's try to optimize these systems, rather than following the conventional wisdom of the industrialized nations — which we know is inappropriate for a world with 6.2 billion people. As my old mentor, agricultural engineer Tod Neubauer said, "It is better to be crudely right than precisely wrong!"

Case Study:
Casas Que Cantan: Community Building in Mexico

Susan Klinker

In 1995 the Obregon office of Fundacion de Apoyo Infantil (FAI, Save the Children) in Sonora, Mexico initiated the construction of a new headquarters that would demonstrate ecological design principles and use labor-intensive rather than capital-intensive building technologies. The group chose straw bale building because it is cost effective (the region has abundant agricultural and labor resources) and would provide maximum insulation against the harsh summer sun. Bill and Athena Steen (of the US-based nonprofit The Canelo Project) acted as technical consultants and volunteer partners in the demonstration building experiment. Local builders and outsiders explored, trained, and shared knowledge as inherent components of the development process. The 5,000-square-foot (465-square-meter) building evolved as an organic process, without an architect, over the course of two years.

CATHERINE WANEK

3-9: *Volunteers work with local villagers to place straw bales for this modest house, built for a total cost of less than US$500.*

Inspired by the project's success, the sister of two of the primary builders of the FAI project decided to build a home with her husband in the local *colonia* (informal settlement) called Xochitl (meaning "flower" in Nauhuatl). Their simple home was admired as one of the most substantial, beautiful, and comfortable homes in the settlement.

An ambitious group of women in the settlement petitioned for support in their efforts to construct similar homes. After unresponsive appeals to local authorities, the group addressed a letter directly to Bill and Athena Steen, inviting them to attend a community meeting.

The Steens agreed to support the group by providing technical support and training but entrusted primary organization and decision making to the group. So the women structured a process of shared work and responsibility: each would help the others build new homes for themselves, beginning with the most needy members.

The Steens provided an integral yet detached role in the project, interjecting much needed moral support, momentum, and technical expertise consistently over time. By linking the Xochitl project with their own nonprofit organization, the Steens provided further opportunities to expand the project across the social, economic, and political boundaries that separate the US and Mexico. The Canelo Project was able to channel donations for Casas Que Cantan (meaning "Houses that Sing") to cover material expenses, especially for foundation and roofing materials (approximately US$500 per house). The organization also conducted a series of educational workshops that included volunteers and students from the US. Participants worked side by side, learning together and engaging in a cross-cultural exchange that resulted in not only the 12 physical structures that have been built to date, but also in friendships, increased understanding, and changed perceptions.

For the Steens, witnessing the spontaneous expansion of natural building ideas in the community has been one of their greatest rewards. In 2001 FAI initiated a new project intended to shift straw bale building toward the private sector, with an ambitious program to build 150 affordable straw bale houses within three years. The project will use technical knowledge gained to date and will not include the Steens or The Canelo Project in any significant role — perhaps the best possible way for participants to demonstrate their self-reliance and local capacity building.

Lessons Learned

1. Strong human relationships are extremely valuable for project development.

Friendships between the Steens and the extended families closest to the building projects in Cuidad Obregon encouraged the growth and development of natural building in the area, a process more spontaneous than calculated. The strength of family-to-family relationships kept the Steens deeply connected to the community and was the root of their joy and sense of reward in their work. Project development grew from and because of friendships.

2. Long-term relationships between a specific organization and a specific community count.

What started as a consultation by the Steens/The Canelo Project for an institutional (quasi-public) building for Save the Children grew to become a two-year building experiment, then shifted to provide technical support in a series of women's cooperative building projects. Straw bale building has now taken on a life of its own in the area. The careful maneuvering of the Steens and their distant, yet consistent relationships with local people have nurtured the use of natural building technologies in Obregon and helped to actualize the ideas and initiatives of the community.

3. Technical solutions developed on-site are often the best solutions.

Every natural building project must be considered within its own ecological, social, and cultural context; and each community offers its own unique combination of local, natural, man-made, and human resources. In Obregon technical decisions were made on-site through a dynamic process of dialog and experimentation that took into account material availability, cost, durability, and ease of construction (simple tools and semi-skilled or unskilled community labor being an important factor).

Certain construction techniques, such as those using the local reed carrizo were specifically developed during the project and were based on modifications of traditional building methods. The development of straw-clay blocks for building was another valuable innovation (see "Straw-Clay Blocks"). Other vernacular techniques were also explored, such as the use of Nopal cactus juices to enhance adhesion and durability of lime plasters. And local knowledge made it possible to select the longest-lasting species of palm for roof thatch and to locate the best clay sources and colorants.

4. A long-term process of experimentation and discovery benefits both the local community and the support organization.

The vigorous, collective exploration of possibilities for natural building allowed new combinations and solutions to emerge. Methods and techniques that became familiar through the process have since become part of the working vocabulary for The Canelo Project in the US, as well as for the project participants in Mexico. Knowledge, skills, and experience gained from the earliest works have been used time and again on both sides of the border. To quote Bill and Athena Steen, "Our worlds, cultural pride, and differences had combined to make something more than a building. Our working and being together had redefined the way we see the world, the way we live, and the way we build. The lives of everyone involved were forever changed"[1]

5. It is better to work with an attitude of respect, support, and mutual cooperation than from a "charity" or "aid" mindset.

The sincerity of this philosophy underlies all Casas Que Cantan work. Workshops are structured as cooperative and educational cross-cultural experiences. Participants come to learn new skills, as well as to support those in need. Everyone gains knowledge, experience, and comes away with new expanded views. Rather than focusing all of their efforts on fundraising and building houses quickly, the Steens have made themselves and their organization available to the community as a steady and dependable resource. Although they assist by seeking material and financial resources on behalf of the community where appropriate, those efforts are limited. Over time their work has slowly grown, changed people's thinking, and served to empower people with skills, confidence, and new options in their lives.

AN UPDATE ON CASAS QUE CANTAN

Bill Steen

The Casas que Cantan project lives on but not quite in the same format as it was started. The original group of women who had committed to one another have finished their obligations and settled back into their lives. Most of their small $500 houses are more or less okay, though many of them took quite a beating in a hurricane in 2001. We have now established a solid core of workers, many of them young boys, who are familiar with straw bale building techniques. They are the backbone of the local housing efforts — without them, none of the work would be possible.

I never cease to be amazed at the continuity of the efforts in Mexico. Our colleagues' efforts never cease, despite the failures and weaknesses that one encounters in developing something outside the norm. Dependency, for example, was a major weakness during our work in the past: whenever there was money for a building and we were there, things happened. However if money, transport, or consistent access to tools were lacking, then things would often grind to a halt.

A potential solution to the dependency problem has opened up through a church that has a branch dedicated to development in foreign countries. Their funding may help those we have worked with to acquire transportation, trailers, tools, and a storage yard/building, allowing them to become contractors rather than low-paid employees. To create a legal entity eligible for grant money, about 15 or 20 builders have formed the Casas que Cantan Cooperative. The church has encouraged them to apply for additional funding that could buy a prickly pear cactus orchard, fencing, and a drip irrigation system.

I would call the Casas Que Cantan project successful, in that more people got trained along the way and we learned a whole lot about building simple, inexpensive shelter. But perhaps our most valuable lesson was that good construction details make all the difference in the world.

STRAW-CLAY BLOCKS

Joseph F. Kennedy, from an interview with Bill Steen

SUSAN KLINKER

In their work in Ciudad Obregon, Mexico (see "Casas Que Cantan"), Bill and Athena Steen have innovated several new systems of construction with their Mexican colleagues. Among the most exciting is a lightweight "straw-clay" block that is used like an adobe (mud brick). The high clay soil of the region results in adobes that crack excessively. Also, although bales of straw were available and appropriate climatically, straw bale construction was unfamiliar to local people. And the bales themselves were susceptible, due to the lack of available tarps, to rain damage while under construction. Straw-clay blocks were an attempt to marry the advantages of adobe and straw bale into a low-cost system that could be widely adopted by local people using native skills.

Early attempts to mix the high clay soil with straw were based on the German-style *leichtlehm* (light-clay) system, but the Steens soon found that the complicated process of ramming the material between forms was not appropriate for the local culture. The Steens recalled some earlier experiments that had not been further pursued, in which blocks were made using straw and clay. They realized that making such lightweight blocks fit within the traditional adobe culture and could be a way to use local materials effectively. The new straw-clay block was first applied to the Save the Children office building in Ciudad Obregon.

3-10: Straw-clay blocks are used to create a vaulted structure. Although similar to traditional adobe bricks, straw-clay blocks have a much higher insulation value.

Advantages

The straw-clay blocks have characteristics halfway between adobes and straw bales. The addition of straw (and experimentally other materials such as sawdust, agave stalks, wood chips, and even pumice) to the high clay soil make the blocks more stable, as well as lighter in weight and easier to work with. The blocks do not need the protection from rain that straw bales do, eliminating the need for tarps — apparently the clay content of the blocks is sufficient to wick water away from the straw so that it doesn't decay. And the straw in the block precluded the erosion or dissolving of the block when exposed to rain, acting as a "thatch" on the surface of the block and effectively shedding water.

Because the blocks were discrete standard units the system lent itself to local enterprise and proved to be a good way for people to make money on larger building projects. During construction of the Save the Children office building, local people — from children to grandparents — contributed to block making and were paid for each block they made.

Although the blocks can be made any size, the smaller blocks proved ideal for interior partition walls and smaller houses. Having less mass than traditional adobes, straw-clay walls do not soak up as much heat, providing a more pleasant interior temperature. The blocks were easily understood by a culture familiar with adobe construction (they are essentially lightweight adobes) and have been adopted much more readily that other building systems, including straw bale. Further because of their weather resistance, the block walls can remain unplastered for awhile before receiving a protective coating of earth or lime plaster.

Disadvantages

Probably the main disadvantage of the blocks is that they have shown some susceptibility to termite damage. This has not been a general observation but a phenomenon restricted to a few buildings, and proper protection of the blocks from the ground easily solves the problem. Another question has to do with the variable density of the blocks. Mostly used as infill for post and beam systems, the blocks could conceivably have load-bearing capacity, though establishing compressive strengths for different densities of mixes is needed before experimentation can begin. The Steens are currently working with Provay (a local nonprofit that has grown mostly out of the Steens' work in Mexico) to further develop the block system. Provay was formed by members of a wealthy family who were inspired by the potential they saw in natural building and decided to pool their resources to produce affordable housing.

STRAW-CLAY BLOCK ROOFS

Bill Steen

As history and modern experience have taught, the roof is the defining element of a building. In our work in Mexico, the most successful roof for simple buildings proved to be a shed roof with corrugated metal and no parapets. However, as wood is not in great supply and termite problems make it an unattractive option, we have pursued woodless options, using straw-clay blocks and vaulting techniques that originated in the Middle East. After the construction of a dome and a Nubian style vault, we learned that most of the recipients of these buildings didn't want domes, and that three-foot-wide (meter-wide) vaults in succession weren't popular either. We also continued to find that lime plaster over clay on these roofs proved too much of a challenge. The most popular roof proved to be nine-foot-wide (three-meter-wide) vaults built out of the local carrizo reeds, but this solution needs a concrete structure to support it and a cement-based covering. Nonetheless we continue to work in the direction of woodless natural roofs and plan to have John Norton, who has been working with woodless construction in Africa (see "Woodless Construction"), help to devise strategies that might be more practical and acceptable in this situation.

In an effort to reduce the amount of wood and/or concrete in the roofs of the buildings they construct, the Steens have been experimenting with using the blocks to build vaults and domes. Inspired by the work of Hassan Fathy (see "Elegant Solutions") as demonstrated by Simone Swan in Texas, the Steens used some rudimentary knowledge they collected on earthen vaults and domes to investigate possibilities. They also contacted a Danish group (AHSA) out of Bolivia that was working with woodless construction. Based on what they learned, the Steens successfully created a prototype caternary vault structure of straw-clay blocks. And although it is promising, the system needs further development to be a viable system for the area. The Steens' experience shows that a particular technology is not static but must respond to structural, cultural, and other aspects of the building process in order to be successfully adopted.

CHAPTER 4

Down-to-Earth Technology Transfer

Kelly Lerner

When I first traveled to Mongolia in 1997 to work on a United Nations Development Program project building straw bale clinics and schools in rural areas, I was terribly naïve about international development. Professionally I specialized in passive solar heating and straw bale construction, and with past experience living in a foreign culture, a degree in sociology, a love of adventure, and a strong stomach I was relatively well-equipped for working in rural Asia. But all my technical and cross-cultural experience didn't prepare me for the challenges of transferring sustainable technologies in a foreign context. Although we successfully constructed straw bale buildings, I usually went home feeling that the construction technique wouldn't be duplicated. If sustainable technology adaptation and adoption was the goal, my early projects missed the mark.

Since then I've spent every summer in northern Asia (Mongolia and China, with a brief project in Argentina in 1999), introducing straw bale construction in rural communities with the Adventist Development and Relief Agency (ADRA). The projects have been both small and large: from designing and building a single seismically resistant straw bale school (the first straw bale building in China; see "Seismic Solutions for Straw Bale Construction") to building 340 straw bale houses in the summer of 2002. I've trained hundreds of construction workers and building professionals and visited construction sites all over northern Asia.

By my own assessment, some of our projects successfully planted straw bale construction in the local communities and some did not. The reasons for success or failure have been multifaceted and different in each case. But we learned from each project, and learning can transform an apparent failure into a success. I still have more questions than answers about how best to transfer sustainable technologies and adapt them to specific communities, but out of my failures and successes I'm developing some general guidelines.

Local Solutions for Local Problems — Go with Questions

An outside "expert" may know a lot about building with natural materials, energy efficiency, permaculture gardening, or some other sustainable technology, but the locals are experts on everything else. They know how people live and work and know about local needs, aspirations, and current issues; climate; environmental concerns (or lack of concern); typical designs for houses or schools; where and how to get materials and how much they cost; how buildings are designed and built and who builds them; who holds the real power; how to navigate local politics and family politics; what the best channels are for education; and where, how, why, and when money flows. Collectively they know everything about their place, and an outsider knows nothing. A lack of local input will assure ultimate project failure.

Questions are the most important tools in sustainable development. So is the ability to listen, analyze, and ask some more questions. Use your nose, ears, eyes, gut, and ask about everything. While you're listening, be aware of the motivations and bias of the speaker. White-collar workers such as government officials, architects, and engineers can provide great information about official systems and regulated building, but they often know very little about vernacular building in villages or what happens at a rural construction site. For developing nitty-gritty construction details and buildable designs, my best resources and design partners have always been site foremen and contractors — the ones who work on-site, get their hands dirty, source materials, know local skilled resources, and plan and supervise the daily work.

When the time isn't appropriate for questions, there's also a lot to be learned by just hanging out, watching, and listening. Important tidbits can fall into your lap while you're sharing a meal or evening libations. The more time you can invest in developing trusting and relaxed relationships, the more information will flow and the more you'll be able to work closely and cooperatively with your partners.

KELLY LERNER

4-1: *Showing examples of successful existing structures is a crucial step in convincing understandably reluctant homebuilders to take the risk of using a new technology.*

The shape, function, and materials of housing are at the heart of culture. A house is a family's physical and financial security. It's probably the largest asset a family has. It's impossible to create a good housing solution (technologically or esthetically) outside of the local context or without local input. Though it may seem difficult or inefficient, developing a solution in partnership with local expertise is the most efficient and effective, if not the only, path to successful technology transfer.

Determining the Goal:
Constructing a Building versus Technology Transfer

Builders are focused on designing and building, usually one building at a time. But it's a big, big world out there. One building by itself, or several buildings, or even hundreds of buildings will have little impact on long-term sustainability or improve the living standard for more than a few.

For real environmental impact and long-term sustainable development, technology transfer must be the goal. That means there will be real and measurable changes in the knowledge, attitudes, and practices of the local community long after the project is completed. Like any other worthwhile project, technology transfer is a slow evolutionary process that requires long-term commitment and an ongoing relationship. Like learning a new language, there are missteps and course corrections along the way. A single building is one good tool in technology transfer, but it is not the primary goal.

I've seen (and worked on) many so-called successful building projects in developing countries using natural materials and sustainable technologies. Communities were interested to see the new building technique and to be involved in the construction. The building technique had been reasonably well adapted to local conditions, and the building was high quality. But far too often after the building is completed, the "experts" and volunteers go home, and the project money stops flowing. Everything returns to the way it was. The building may be exquisite (though usually it's not) and the community may benefit from its use, but they aren't using the natural building technique that they learned.

Don't get me wrong — constructing buildings does play a fundamental role in introducing sustainable building technologies:

- They are testing grounds for new techniques and materials.
- They provide real, local, "kick the tires" examples.
- They provide learning and training opportunities.
- They build relationships.
- They are real-life laboratories for monitoring and feedback.

AN UPDATED SAYING

- Give a woman a fish and she will eat for a day.
- Give her a pole and teach her to fish, and she will eat until the pole breaks or the pond dries up.
- Teach her to find the materials for the pole, line, and hook; how to build a fishing pole, locate good fishing holes, and fish; and she will eat for the rest of her life.
- Give her a micro-loan and teach her to manufacture and sell fishing poles and set up a fishing training program, and fish will become the new staple food. Over-fishing may cause an ecological disaster (Yikes! Unexpected results. Remember to monitor).
- Give her a micro-loan and teach her to raise fish in an ecologically balanced way (feeding them waste products and selling both fish and fish fertilizer), and she will adapt and optimize the system through trial and error. All her neighbors will eat well, see her success, and start their own fish-farming operations. The whole region will thrive (and the World Bank will come study your program for possible replication in other regions...).

And there are many good ancillary benefits to constructing even one building:

- Cross-cultural friendships and broadened world views for all involved.
- Real physical and financial benefits to the people using the building.
- New skills for the lay and professional builders.
- A needed influx of energy or funding for a worthwhile project.

But what happens after a building is finished? Does the community have all the skills they need to go forward with this new building method — planning, researching (often from resource materials in a foreign language), modifying, designing, engineering, gaining approvals, financing, locating tools and materials, and building? If the goal is technology adaptation and transfer, we must broaden our focus beyond simply teaching natural building techniques and building demonstration buildings.

The development of a newly introduced sustainable technology in a developing country is not so different from a similar process in the developed world. For example it has taken years of concerted effort and strategic forethought by a multitude of devoted straw bale supporters to move straw bale construction from obscurity to its current eco-groovy fringe success. Why should we expect anything different elsewhere?

We don't always have the good fortune of a project with long-term funding focused on technology transfer. Does this mean that we should never do "one-off" demonstration buildings? It all depends. Clearly there is value in helping an individual or a community with a building, and value in the broadened cross-cultural understanding that can come by working with others. But can the project be organized as if it were going to be the first chapter in a long running relationship? Can the focus be on education by way of building? If the answer to these questions is "No", I'd go elsewhere. There are many needs, and our time is the most important limited resource we'll ever have.

Marketing: Selecting Target Populations and Building Types

Everyone feels good about helping the poorest of the poor — the homeless, the disabled, the single parent family — and they definitely need assistance. Keep in mind though, if technology transfer is the ultimate goal, building for the very poor may not be the best entry point for a sustainable technology. In fact building primarily for the poor may relegate the sustainable building technology to the category of poverty housing and make it less appealing to the general population.

Every part of society looks up to the one above it. What are people living in; what products are they using; what do their houses look like? I've often thought that a beautiful, well-publicized straw bale house for Madonna, Tom Cruise, or Michael Jordan would advance straw bale construction more than all our previous efforts put together. Part of technology transfer is marketing — expanding awareness about the benefits of the product or process, making it as appealing as possible, and ultimately creating consumer demand. Schools, clinics, restaurants, and other public buildings provide a great opportunity to showcase sustainable technologies to a large and ongoing audience. Because the budget is often larger for public buildings, there are more opportunities to try new techniques and focus on fine finish details.

If the materials themselves are affordable and the construction technique itself is kept simple and clear, individuals will see the potential, and the technique will filter down. The Casas Que Cantan project was initiated by poor women who had witnessed the construction of a large straw bale building (see "Casas Que Cantan"). They had experienced straw bale construction on a large, nice building without any risk to themselves. No one had to tell them it would work on a smaller scale. In Mongolia several builders who trained on straw bale clinics went home and built small, simple straw bale greenhouses for themselves.

Although building a public building can provide exposure for a sustainable technique, it's better to start with smaller buildings in order to become familiar with local conditions and work out adaptations. Be especially careful of media exposure in the early stages while research and development are the focus. Although it's great to have lots of coverage with a successful project, the media can be brutal about mistakes and missteps.

If the goal is to introduce a building technique for houses, then it pays to choose the target market wisely. In China we've focused our straw bale housing efforts on the rural "middle class" — farmers who were already saving money to build a house on their own. As some participants are relatively well off and some are quite poor, some of the finished houses are quite nice and some are just so-so. Potential builders and owners see a wide spectrum of what's possible with straw bale.

Rigor and Excellence: Build the Best Quality that LocalMaterials, Local Skills, and Budget Will Allow

As I train and build in a new community, I try to remember that this is the only example of straw bale construction that these folks have ever seen. Even if I show lots of pictures, their main impression of straw bale construction will be based on the quality and performance of this local building. If the building is difficult to build or isn't warm, or if the walls rot after the first rainy season, people will assume that all straw bale buildings are difficult to build, aren't warm, or aren't durable. The stakes are high, and the last thing I want to do is build a low quality example of a new technology I am trying to introduce.

Although there will certainly be missteps and adaptations along the way, the transfer of technology requires a rigorous approach — a commitment to providing the best quality example (judged by local standards) possible.

Purism or Practicality

By training and natural inclination, builders, architects, engineers, and scientists generally focus on the three-dimensional, material world. What does it look like? What is it made of? What's its function? How does it go together? How does it perform? In their quest to design an elegant and effective sustainable solution, they often lose sight of the practical considerations of local implementation — especially if they are working outside of the local context.

In order to be effective sustainable technologies must be adopted and widely used. To be adopted they must be easy to use and fit well with existing local materials, technologies, and skills. Efficacy, not purity, ultimately transfers the technology.

For example, currently in China our hybrid building systems use:

- Rock foundations (with or without a grade beam depending on seismic activity and local practice).
- Full-width brick columns, flanking window and door openings.
- Straw bale in-fill between posts.
- A concrete bond beam.
- Locally produced trusses.

KELLY LERNER

4-2: Mud is used to stabilize relatively soft bales.

Tile or Metal Roofs (Depending on Local Practice)

Although the finished houses are 68 percent more energy efficient than the standard brick construction (in terms of thermal bridging and material use), this construction system is less than ideal. A "pure" load-bearing straw bale system would provide better insulation and less embodied energy.

But load-bearing straw bale has other drawbacks. Load-bearing straw bale walls don't provide enough support for heavy tile roofs when combined with large south windows (for passive solar heating). Local windows are designed for installation in brick walls. The wood required for rough bucks in a load-bearing wall is expensive, and the carpentry tools and skills are rare. The rainy season coincides perfectly with the building season, and the tarps needed to protect load-bearing walls through a heavy rain are expensive and of low quality.

So, in spite of the energy-efficiency drawbacks and added embodied energy of brick, given all the variables and worker's skills in the local context, a brick post and beam in-filled with bales works better than any other system we've explored so far. Most importantly the local builders and owners understand and like the system and are using it.

As straw bale building continues and different materials become available, we are constantly improving the system — introducing more insulated elements (coal slag under slab floors, thermal breaks in the concrete sills, rigid insulation in the bond beam), teaching builders how to formulate good plasters with local materials, teaching local project teams some management skills, and improving our training and own management approaches.

In design and construction I'm always balancing actual, real-life conditions with design ideals. In my experience it's better to have a "bastardized" system that actually works and is easily replicated than a "pure" system that dies after a few buildings because it is too cumbersome or doesn't fit well with local materials or skills. Only if buildings are actually getting built can we get feedback and keep working to better them.

In many arenas developing countries are technologically leapfrogging developed countries because they aren't encumbered by an existing infrastructure devoted to an outdated, inefficient technology: cell phones abound in even the most remote corners of China. I fully expect that sustainable photovoltaic and fuel cell technologies will take off in the developing world long before taking hold in Europe and the US. Energy-efficient design and natural building techniques could enjoy the same success, but they must be esthetically and technologically adaptable.

Many in the developed world, nearly smothered by their material and technological riches, are beginning to evaluate their true well-being. They feel the negative results of living with industrially produced, synthetic materials and long for a slower, simpler lifestyle in balance with the cycles of nature. They discover the joys of natural building and start with a curvy cob garden

4-3: *Cultural acceptability of an innovation is crucial if it is to be adopted. Although this straw bale vault in Mongolia is technologically advantageous (it is highly insulating in a country where a high proportion of income in winter is used for fuel), the shape has proven to be culturally unacceptable. This particular straw bale system has not been replicated.*

4-4: *A hybrid technique that uses a brick post and beam system with straw bales as infill has proven to be successful in China. This system evolved as a result of constant feedback from local builders and attention to local stylistic preferences.*

shed with thatched roof, then move on to build a passively solar-heated, straw bale house with rammed earth interior thermal mass walls, and all earth and lime plasters. They can wax poetic for hours on the joys of irregularly curving bale walls plastered with earth, and their subtle textures.

The view from a 40-year-old crumbling earth and rubble house with a thatch roof in a Chinese village looks a little different. These farmers are definitely in touch with nature — they work long days in the fields or their greenhouses, eat from their own garden, fill their 30-gallon water jars twice a week when the water is on, empty their own pit toilets, and tend and butcher their own animals. There's enough to eat and the children go to school, but there's little cash for extras or emergencies.

KELLY LERNER

Villagers watch their urban peers on TV and long for hot and cold running water, indoor bathing and toilets, tile floors, big windows, refrigeration, telephones, a roof that doesn't leak or need repair every year, walls that wouldn't melt slightly at each rain, and machines and tools to relieve tedious physical labor — and a new, larger TV. They're saving money and stockpiling materials to build their son a new house so he will attract a good wife and take care of them in their old age. The new house will be substantial and durable brick, with straight flat walls, a white glazed tile exterior, a red tile roof, blue-tinted windows, running water and electricity — just like the village leader's house down the lane.

4-5: A conventional brick house with metal roof in China. Although this house uses desirable industrial materials (bricks, concrete, and a metal roof), it is hot in summer and cold in winter. It is ironic that the straw next to the house could have been used to improve thermal performance.

If a brick post and beam, straw bale infill house can fulfill this dream for a durable, substantial, straight-walled, sharp-cornered plastered brick house at the same cost *and* use less fuel for heating *and* be more comfortable in the winter, then it might catch their interest. A little financial incentive, training, technical support, and some high quality examples might even persuade them it's a good investment of their whole life savings for their family's future.

After careful consideration they build a straw bale house. With an attached greenhouse, it's a near duplicate of its neighbors, but after the first winter they love it as their own. They glow with pride in front of their white tile exteriors and can wax poetic for hours on the joys of seeing their arthritic crippled mother walk around even in the coldest months and their baby stay healthy all winter.

Straw bale construction in China is growing, primarily because a plastered straw bale house can look identical to the highly esteemed brick house and cost just about the same. It also performs much better, but energy efficiency wouldn't matter much if the straw bale house didn't look right esthetically (straight walls and right angles).

Natural sustainable building won't go forward if it's yoked to a particular style. Whether working at home or overseas, I can't assume that my esthetic values about natural building are superior to someone's dreams for their home. It's their house. They have to live in it; they have to love it. I can go home and build my own house.

4-6: *A completed brick and straw bale house. Note its passive solar orientation, lime-plastered walls, and conventional roof.*

Take the Long View: Innovate Incrementally

Building is an intricate system with innumerable parts — design, engineering, sourcing materials, windows, doors, auxiliary systems (water, electricity and heating), labor and supervision, financing, etc. Just as with plants in nature, the best construction systems develop organically over time in concert with climate, locally found materials, and the skills of the local builders. When introducing a new technology, rather than spend energy trying to reinvent the wheel, start by studying and adapting existing local technologies.

In the US lightweight post and beam or modified wood frame systems in-filled with bales have made straw bale construction more understandable and appealing to builders, engineers, and code officials. In China (with its abundance of bricks and masons) brick posts and concrete beams in-filled with bales fit well. In Mexico (with its tradition of adobe block) the idea of light straw-clay blocks transferred easily.

The first stage is research and development — small, low-profile projects acting as good low-risk experiments. A small project, with lower costs and lower energy investments, can explore different options quickly without high financial or political risk. Neither you nor your local partner knows just how a technique will work in the local context. Building it in three dimensions will quickly point out both suitability and problems, as well as allow for modifications — in materials, construction technique, or management. Later projects can be built on the information gathered from early projects.

ADRA started with one school in China in 1998, and the projects grew incrementally as we gained experience and the support of local communities: 21 houses in 1999; 75 houses in 2000; 170 houses in 2001; 340 houses in 2002; and 490 houses expected in 2003. The mistakes and successes of early projects support our current work, and the work this year teaches us how to do it better next year.

Monitoring, Evaluation and Innovation

Project evaluation is an essential aspect of ongoing technology transfer and adaptation. How will we know what changes could be made without a careful evaluation of project strengths and weaknesses?

QUESTIONS TO USE FOR EVALUATING A DESIGN

For the owners/occupants:

- What do you like best about this house?
- What do you like least about this house?
- What changes would you make in the building, if you had the chance?
- What heating system do you use? What fuels do you burn? How much and how many times per day? Do you use the same fuel for heating and cooking? How much do you spend on fuel for a year?
- Is the temperature inside comfortable in the winter? In the summer? How many layers of clothing do you wear inside in the winter?
- Where are the coldest and warmest parts of the house?
- How much does the indoor temperature fluctuate from day to night?
- Have you noticed any changes in your family's health since you moved into this house?

For builders (owners or contractors):

- Did you receive training? How much and from whom? Do you feel comfortable with the technique?
- Was it easy or difficult to build?
- What were the most difficult parts of the construction?
- What changes and improvements did you make?
- Were all the materials easily available? Which materials were most expensive?
- Was the technology easily taught?
- Do local builders understand critical health and safety components (waterproofing, structure, fire resistance, full insulation) enough to avoid fatal mistakes?
- Would you build one for your own family?

Even if the construction is successful, what are the long-term effects on the family and the larger community? What are the unintended effects? Consider the example of the Nepalese village, where a very successful photovoltaic electrification project led to local deforestation. Because they now had electric light, families were staying up later at night, needed more heat, burned more wood, and cut down more trees. Projects fail and succeed for many different reasons. In order to see the whole story, evaluations must consider technological, management, financial, and cultural/sociological issues.

Challenges may be technological: plasters cracks; windows are difficult to frame; the heating system is too large given the level of insulation, etc. There may be cultural or esthetic barriers: the foundation isn't high enough; the windows aren't large enough; the roofing is the wrong material; construction is more expensive than standard construction techniques; or the house just doesn't fit in.

But often project problems relate more to management and delivery: project funding isn't flowing smoothly during the building season; the wrong people come to the training (managers instead of builders); contractors try to cut corners; conflicts and infighting occur between local government agencies; there are poor communications or lack of connection between urban managers and rural implementation; the right building materials (bales, for example) weren't secured in advance; new inhabitants don't receive training about the new building operation and maintenance, etc.

When trying to evaluate a project, it can be very difficult for a foreigner to get honest answers. Cultural imperatives about hospitality may make it difficult to share problems. And when high-ranking officials tag along on site visits, inhabitants may feel obliged to paint the best possible picture of the project (to protect their own interests). In China I often visit houses with the unshakable management entourage and chat with owners just to get an overall impression. My translator then comes back later alone for informal conversations, where she follows up on questionable items.

Multi-Level Education

A newly introduced technology can face many barriers from many different levels: owners, local government, builders, designers, engineers, material suppliers, provincial government, regulatory agencies, national government, etc. For everything to go smoothly each level must be convinced that the new technology benefits them in some way.

Owners need to see how the new type of building can make their life better: Is it easier to build or less expensive or warmer or less expensive to operate or healthier? Local governments need to see how the technology fits into its plans and goals and what benefits it brings to the local economy or local services. Regulatory agencies need to know that the technique is safe and how to evaluate it. Builders, designers, and engineers need to understand how the whole building goes together and how to work with the materials. National governments must see the value in the project and support the local governments.

It is imperative to provide education about the sustainable technique to as many levels as possible and to adapt the message to the concerns of the audience. Owners need information about energy efficiency, fuel use, and maintenance; builders and designers need to know how to work with the materials. Good contacts in the national and provincial governments may provide invaluable guidance about working in local areas and introductions to allies. A good working relationship with local leadership can be critical to project success.

WHO BUILDS?

Although using a group of volunteers from the developed world to build a house or clinic in a developing country can be meaningful, rewarding, and great fun, I struggle with this approach to sustainable development. On the one hand, poor communities can benefit from an injection of energy. And cross-cultural friendship and understanding are essential to understanding our role as world citizens on a planet with finite resources. Living with less and still enjoying themselves can motivate volunteers from the developed world to better understand and act on their responsibility to reduce their own resource use.

On the other hand, it takes a lot of money and resources to send all those volunteers to a rural site. Their needs for food, lodging, and facilities can overtax local facilities and hospitality, and foreigners often fall sick from local food and water. They don't know local materials and construction systems, and they don't speak the local language. Is this really an effective approach?

If technology transfer and adaptation is the goal, wouldn't the money be better spent in the local community, training and paying local builders who already have lodging and can eat their native food? The US$900 spent on a round-trip airfare to send one person to China could pay the wages of an entire local work crew to build a 600-square-foot (56-square-meter) house.

If outsiders come, simply build a house, and then leave, then the technology leaves with them. If local builders learn the new technology from design conception to completion, it can become their own. They become empowered to adapt the technology and apply it in ways that fit.

When a group of outsiders (with their fleece jackets, radios, and cool pocket knives) is transplanted into a rural community, there's always the risk of cultural imperialism ▶

— the incorrect but prevalent view that the materially rich, dominant "developed" culture is better than impoverished, backward "developing" culture. Though volunteers may not have this attitude, the rural community may start deferring to them — asking the "experts" everything, rather than relying on and developing their own skills.

Though I love spending time in rural China and have learned greatly from the people there, I've found that technology transfers most effectively if keep my time in rural villages short. I train as well as I can, work with locals to fine-tune their design, and then get out so they can figure things out on their own. They know I'm available for technical assistance and that we'll be making periodic site visits to inspect and advise, but they also know that when they ask a question, I'll ask them their ideas for possible solutions and help them choose one.

Equal Investment: Everyone Pays, Everyone Plays

When all project partners from the bottom to the top — homeowner to highest government official — become true stakeholders and are equally invested (emotionally and financially), everyone has the same strong motivation to successfully accomplish the project and keep costs down. Since the stakeholders are full participants in the planning, design, and construction, they feel a true sense of ownership — for the project and finished buildings.

In China our housing projects are structured so that ADRA pays a fixed subsidy (about a third of the construction cost), provides training, management, and technical support. The local government pays a third and provides a local management team. The new house owners pay the remaining third. The contributions are often in-kind rather than straight cash — sometimes the local government provides materials at cost that they've purchased in large quantities at a discount, or they may waive land costs or other fees. If the owner has building skills, they often provide labor, thus reducing the overall cost.

Effective Training: Teaching Each Other and Empowering Local Building Professionals

Standing up in front of each new group, I'm struck by the fact that though I have lots of experience with straw bale construction, my "students" have *all* the information about local building materials and systems. I have more to learn than I do to teach. And the best solutions always come out of dialogs with local builders. I don't teach as much as I facilitate and guide a design process. By the end of a weeklong training, the local team (made up of a project manager and builders) will have designed their very own straw bale building, adapted to their local circumstances. If a picture is worth a thousand words, hands-on training must equal a whole library.

Although the classroom presentations only require one person, it's very helpful to have several skilled builders reviewing the assignments each day. If the training site is adjacent to an in-process construction site, it would be better to format the class with a half-day hands-on, and a half-day in the classroom.

MY OWN GUIDELINES FOR NATURAL BUILDING PROJECTS:

- Go only where there is locally expressed interest and grassroots support. There's no point teaching if the students are not interested. Though their motivations may be different from my own, I expect my students to be at least as enthused as I am about their project. I also expect them to have made some financial or time commitment to the project.
- Give adequate time to the early phases of research, training, cooperative design, and planning. Good planning can make or break a project. Building well takes time and resources — research, planning, and design are the least expensive parts of the process. Scrimp on the early phases, and the later phases are bound to suffer.
- Don't let funding dictate your project timeline. (This is a common pitfall, almost inherent in the funding system). The project proposal is written by a talented and visionary development professional who knows a lot about how to raise money, but little about the ins and out of construction: It's a long way from project proposal to finished project. Unexpected roadblocks often appear — torrential storms, alcoholic personnel, electrical blackouts, political gridlock, labor strikes, national holidays, material shortages, cash flow delays. Building projects are constrained by the building season. At some point, if the design can't be worked out or the personnel isn't in place or the materials aren't available or normal setbacks have made it too late in the building season, you don't want to be forced to go forward and build a specific number of houses rather than wait until all the preparations are in place. Some might argue that the time is never right and sometimes you just have to move forward. But to cite another example, it's unwise to start building late in the season, plaster when it's freezing, and then complain that straw bale construction has problems with cracking plaster.
- Try to write proposals with timelines that allow for adequate research and preparation and allow for unforeseen circumstances like bad weather. Always ask: Will going forward at this point help or hinder this technology transfer?
- Couple good technical training with local design. The science of natural building is still young, but we do know a lot about what works and what has failed. I try to provide my students with the most thorough and up-to-date information possible (appropriate to their climate and building systems). We then work together to adapt natural materials to local designs and local systems.
- Keep sight of the ultimate goal — technology transfer. Ultimately my primary goal is to develop local experts who can take the work forward on their own. I want to work myself out of a job, so that I can go to some other interesting place and do it all over again. Or better yet stay home, garden, and build myself a straw bale house.

Case Study:
Zopilote: Traditional, Colonial, and Ecological Building in Mexico

Susan Klinker

Zopilote is a small locally based environmental organization in Tlaxco, Mexico. At least twice per year the organization hosts a two-week study course in the highlands east of Mexico City. Titled *Traditional, Colonial, and Ecological Building in Mexico*, the workshops provide an intensive blend of theory, dialog, and hands-on exploration of natural building technologies and land restoration practices. The unique setting and history of the organization's activities in the Tlaxco region and local community provide the foundation for the workshop's technical presentations, demonstrations, and field trips to relevant regional sites.

The Zopilote course is intended for architects, development workers, regional planners, builders of all kinds, historians, teachers, writers, contractors, and individuals who plan to build their own homes. Most importantly the workshop provides a setting for people with shared interests and goals to build relationships across cultural, economic, and political boundaries while encountering firsthand the rich history and traditions of the Mexican Highlands.

SUSAN KLINKER

4-7: *Paco Gomez lectures to students hailing from all parts of the Americas.*

The Zopilote organization actively promotes sustainable rural development based on the principles of self-reliance and simple lifestyles lived in harmony with nature. That process requires a fundamental shift in values away from Western products and lifestyles (which prioritize economic means in nearly all aspects of life) toward the creation of healthy, contemporary, subsistence lifestyle households and communities. To that end people learn small-scale organic farming techniques, agro-forestry, solar energy utilization, rainwater collection, ecological building, and sanitary composting. In practical application, dependency on outside resources is reduced, and the negative impacts of economic market fluctuations on the community are minimized.

Zopilote workshops feature reciprocal learning, and all participants are encouraged to share their knowledge and experience with the group. The dynamic and open group exchange is guided through a full agenda of topics each day by program directors Alejandra Caballero and Professor Paco Gomez. Ianto Evans of the Cob Cottage Company co-directs, adding specific expertise in popular natural building technologies from the US. Each session maintains a group balance comprising one-third US or

European participants and two-thirds Mexican and Latin American participants, from a range of regional and professional backgrounds. Presentations and discussions are fully translated into both Spanish and English.

According to Zopilote, the course is offered to help accomplish the following objectives:

- To increase the number of people who understand the principles of ecological interdependence and regenerative systems, particularly as they relate to rural development
- To help participants develop skills that can be used to select appropriate technologies and to design /maintain sustainable systems for meeting basic human needs (food, shelter, energy, water, cash income, etc.)
- To help participants become more effective teachers, able to communicate sustainable systems concepts to people of all ages in a variety of situations
- To integrate aspects of each participant's relevant personal experiences, cultural background, and technical skills into the course
- To highlight the experiences of people who live in a poly-cultural society with meso-American roots and ancestral traditions
- To provide another focus, rather than the traditional Western one of "saving the Third World."

4-8: *By visiting traditional earthen buildings in the region near Zopilote, students gain a sense of history, as well as the technological information necessary for developing innovations.*

The Zopilote course thus provides an arena for participants to solve common problems together, while strengthening relationships between people from differing backgrounds.[1]

The organization also supports an accredited alternative primary school that focuses on environmental education and "nurturing a vision for an *"educacion para quedarse"* (meaning "education in order to remain in one's community").[2] The school stresses local activities, the use of local resources, and "invites the child to consider his/her development relative to the local environment, instead of one that breaks ties with their place of origin."[3] Academic work at all age levels is integrated with practical training in carpentry, pottery, textiles, horticulture, and forestry.

The Cob Cottage Company in Oregon, US is a key partner in the Zopilote organization, promoting the workshops, helping to coordinate arrangements among non-Mexican participants, and campaigning for financial support for the primary school (primarily to pay for teachers' salaries). Workshop tuition collected from participants from industrialized nations goes toward scholarships for Latin American participants with fewer cash resources. Needy participants from wealthy nations can do a "work trade" and extend their learning experiences by contributing to the organization's efforts after the course is complete.

Lessons Learned

1. The ongoing process of reciprocal learning and information exchange across cultures is an important element in sustainable development.

First, on a psychological level, the workshop reinforces the commonality of the struggles faced in creating a sustainable future. It comforts people to know that others are dealing with the same issues, and promoting similar ideas in other corners of the world. It gives them hope that even small steps made in the right direction can provide a quiet example, and can eventually have some impact on the greater global society. Participants are also inspired by their exploration of a wide breadth of possibilities in technical solutions, and gain confidence in their abilities.

Second, on a social level, the workshop brings together people of completely different backgrounds and lifestyles and places them in an environment of mutual learning and exchange. Factors relative to social, economic, or educational status that may present barriers to communication in other settings are not apparent. Participants eat, sleep, work, and learn together. All participants including the host/facilitators are perceived as equals, and each person's contributions are valued. Participants come away with new friendships, a network of contacts, an increased appreciation for the way others live, and an expanded worldview.

Finally, on a technical level, the workshop environment creates fertile ground for experimentation and the cross-fertilization of ideas that may lead to new and exciting technical solutions.

2. The seminar format fosters empowerment and individual capacity building.

The workshop approach creates a setting for students to explore and discover new possibilities. As learning occurs most smoothly when students discover things for themselves, many new concepts and a great deal of information can be absorbed within a short period of time. The process targets participants' thinking regarding how we relate to the land as the source of all our basic needs. Although the process is geared toward learning, it is more oriented to exploration than to technical transfer. Consequently participants leave with an increased capacity to problem solve and to apply knowledge in new and creative ways when they return home.

3. A limited scope of involvement can be highly effective in development efforts.

Zopilote has defined its role and built a strong identity around a specific and limited mission. Although Zopilote acts as a catalyst for initiating work in many different regions and social settings, it maintains a manageable and steadfast focus on ecological education. Except for seminars, the organization places limited efforts on increasing public awareness or on changing societal values: they do not publish, collect technical data, or actively network with other national or international organizations. Nor does it retain responsibility for project implementation beyond the seminar context. Instead, Zopilote functions effectively as a resource for information and a catalyst for activism.

SEISMIC SOLUTIONS FOR STRAW BALE CONSTRUCTION

Kelly Lerner

Large faults crisscross northern Asia like a crazy quilt, and straw bale projects are encouraging the development of low-tech approaches for earthquake-prone areas. Bales, with their relatively lightweight and wide footprint, tend to form stable wall assemblies. Most of us designing with straw bales for seismic loads have used conventional steel straps or stucco skins to brace wooden frames (with infill bales), and know that bales provide a good back-up system in case of large earthquakes.

In my home state of California, we have long hypothesized about the ability of straw bales to resist large seismic forces. In 1997 David Mar (a structural engineer at Tipping Mar and Associates, Berkeley, California) proposed a new approach to the structural design and analysis of straw bale walls. His externally mesh-reinforced walls were based on a ductile straw core trapped within a wire mesh basket. With this system special mesh end cleats prevent sliding shear failures, while well-anchored mesh panels provide both vertical and horizontal reinforcement (similar to reinforced concrete shear walls). Under large seismic forces the trapped straw core forms a diagonal compression strut as the wall deforms. The deforming straw core can absorb the seismic energy well after the external skins are damaged.

David was able to test his caged-bale theories in a structural test of a straw bale vault. It performed even better than anticipated: the vault resisted a lateral load 126 percent of its own weight (four times the code requirement for the worst seismic zone) and would not collapse! Most interestingly, although the stucco shell cracked and distorted due to concentrated loading early in the test, the vault continued to resist increased loading even without the stucco skin intact. Those results imply that the system could perform with lower strength lime or earthen plaster, or even mesh alone.

In 1998 I invited David to help with the structural design of four tiny prototype houses in Mongolia. There, extreme cold and limited construction materials control the design, but seismic risks are also significant. Wire mesh wasn't available so we used vertical and horizontal heavy wire ties to contain the bale walls. Because Mongolian two-string bales are soft (even after manual recompression), I stiffened and strengthened the walls with a vertical paired-rib exoskeleton tied through the bale wall. This system integrated well with the wire ties.

Connections at the foundations were especially important to resist sliding shear and carry lateral forces down into the ground. Short wire ties were installed in the foundations with loops on each end. In concrete foundations, wire ties looped under the steel reinforcing and were cast in place. In gravel-bag foundations, wire ties looped under a wooden rib below the top course of bags. The ribs spread the load and prevented the wire from simply slicing through the bag.

Because wire mesh wasn't available, we used wooden lath as the plaster substrate — a choice that degrades the system somewhat, because the wooden lath separates the stucco skin from the straw bales and the wire ties. I still feel confident in the design though: as the vault test indicated, the wire-tie containment engages the energy-absorbing abilities of the bales even without the full strength of the stucco skin. And the paired-rib exoskeleton helps contain and consolidate the wall.

As I traveled China in the wake of a major earthquake, my enthusiasm about applying the caged-bale system kept increasing. The earthquake damage to existing buildings was severe: timbers supporting tile roofs collapsed when existing earth and rock walls crumbled. Though only 100 people lost their lives, the quake leveled whole farming villages. In some cases it was easier to relocate and salvage materials from the rubble of villages than to rebuild in the same location.

The reconstruction effort featured beautiful new two- and three- room, unreinforced red-brick houses, with dry-stacked rock foundations and large timber beams

KELLY LERNER

4-9: *As the preferred wire mesh was unavailable for this two-story straw bale building in Mongolia, sacrificial wooden lath was used to provide a plaster key and additional stability in earthquakes.*

supporting clay and tile roofs. I questioned my hosts about the seeming lack of seismic precautions, and they readily agreed that another 6.2 earthquake would likely destroy these homes, as well — but they had few choices of construction materials, a limited budget, and masonry walls were the vernacular tradition in this area. I was determined to address this problem.

Back in California and working with David, I designed a simple system based on Chinese vernacular architecture and available materials. Given low density two-string bales, lack of wood, and locally produced steel trusses, we decided to use a steel frame structure laterally braced with caged-bale walls.

The mesh-caged straw bale walls provided the most challenge, both in design and execution. In this design a continuous mesh "horseshoe" is cast into the foundation under the horizontal steel reinforcing. The mesh extends above the foundation and ties to mesh panels on the walls. The mesh panels themselves are tied through the wall (inside to out) with heavy wire and a bamboo, paired-rib exoskeleton. Wall ends (corners, T- intersections, and window and door openings) are specially reinforced with continuous horseshoes of mesh and ribs. The ceiling diaphragm mesh also interweaves with the walls, both inside and out. The resulting straw bale structure provides not only energy efficiency and thermal comfort but also security in case of a major shake.

David Mar has continued to move forward in California with this wire mesh-cage structural system, and both cement stucco and earth plaster models are currently (2003) undergoing structural tests sponsored by the Environmental Building Network. If their results prove to be as promising as initial test results, then the prospects are good for promoting the system, not only in China but also in seismically active areas worldwide.

Case Study:
La Caravana: Mobile Training for Eco-Development in Latin America

Alejandra Liora Adler

La Caravana is a mobile ecovillage and training center that has been traveling in Central and South America since July 1996. Staffed by an international group of approximately 20 volunteer social activists and artists, La Caravana inspires and teaches people to take better care of the Earth. Through workshops, courses, theatre, conferences, audio-visuals and special events, we have touched the hearts and minds of tens of thousands of Latin Americans in cities, towns, villages, and rural communities.

Functioning as both a demonstration model of community living and an ecovillage training center, La Caravana has offered courses in permaculture, natural building techniques, appropriate technology, and consensus decision making, as well as many artistic workshops. Volunteer members and in-country and foreign instructors offer courses that, by design, incorporate reciprocal learning strategies. Local populations and project members share both skills and knowledge while on La Caravana. International volunteers then return to their home countries to replicate their learning.

· 4-10: *The La Caravana bus passes Guatemalan children on its ongoing journey through Central and South America.*

Each member of the crew brings his/her particular desires, expectations, limitations, and talent to the community. Since we come from different nationalities and educational and cultural backgrounds, La Caravana provides a way for each of us to confront ourselves through daily life activities, collective work, and the difficulties of travel. This triple challenge brings out the strengths and weaknesses of each member and challenges the group as a whole to overcome difficulties and move forward with its vision and commitment. Over the years more than 350 adventurers (ranging in age from 3 to 68 years old) with the desire to experience a new form of living have served as volunteers on La Caravana. At times our unique community has included a mobile academic school for children and adolescents.

We travel with our own infrastructure of two converted school buses (equipped with solar panels), a pickup truck, and a trailer for our 500-person circus tent. A built-in kitchen and office, specialized equipment for workshops and multimedia theatre activities, a vast array of tools, a library, a large selection of video documentaries, camping gear and each volunteer's personal belongings round out the equipment.

Arriving at a new base we set up a model camp, attempting to demonstrate sustainability and simple living in action. We build a recycling station, a sprouting station, a water-conserving dishwashing station, a composter, and a rainwater collection system — all made from natural, local, or recycled building materials. We cook our own food, have a communal economy for all basic expenses, and make our decisions by consensus. Group work is shared on a rotating basis and volunteers take responsibility for leadership in the various aspects of community life and work.

Using circus and theater arts as its principal tools for approaching communities, La Caravana enjoys an open welcome not usually experienced by tourists and other visitors. As cultural ambassadors we of La Caravana take from each country we visit elements of its history, legends, music, spiritual principles, and prophecies. Through its multimedia spectacle, La Caravana unifies the best of the past and present and offers messages of hope and a planetary consciousness for future generations.

We continue with workshops, (often relating to the arts) and especially involve women and young people. The workshops connect us with the community and give us the opportunity to learn more about people's needs, values, history, and vision. Only then do we begin to plan (together with the community) the longer courses, which often include a workshop in natural building. Depending on the community we may have more to learn than to teach!

Funded in large part by its work for governments and institutions in the countries to which we travel, La Caravana has also received donations from foundations and individuals. In indigenous and other rural communities, we often trade food for work. Scholarships are provided at each workshop to enable local people to participate at reduced or no fees. Often government or foundation support has provided economic aid for participants, and we have received various forms of support from individuals; communities; local, national, and international institutions; and from widespread media coverage.

Recognized as a pioneer model of mobile ecovillages, a school of life, and a training center for future social and cultural leaders, La Caravana represents the mobile arm of the Ecovillage Network of the Americas; forms part of the Global Ecovillage Network (visit us at: www.ecovillage.org); and completes a mission of diffusion, education, sensibility, capacity building, and networking throughout Latin America. Its rainbow flag is a symbol of diversity and the union of all the colors and creatures of the Earth, as well as the hope and opportunity that can arise even amidst social and natural disasters.

A Sampler of La Caravana's Work

An Introduction to Permaculture

Where: El Pauji, Gran Sabana, Venezuela

When: January 1999

Who: Twenty-five Venezuelan and Colombian participants and 15 international participants from La Caravana formed a multicultural group and spent a week together at an arts and community center in the mountains of the Gran Sabana.

What: Courses in sustainable design, permaculture practice, natural building, and appropriate technology. Hands-on technique blended with theory and dialog. The workshop was videotaped and has since been used to teach both in Venezuela and on La Caravana.

Several participants continued their studies with members of La Caravana who specialize in cob construction and began to teach natural building techniques (particularly cob construction) throughout the country. They have since taught dozens of workshops to Venezuelan and international students. Some participants formed an ecovillage network, and the first "Gathering of Ecovillages" took place in December 2000. More than 30 people involved in communal projects and sustainable lifestyles attended.

How: A grant from the Gaia Trust enabled us to offer the Introduction to Permaculture course at reduced fees and to bring instructors from the US and Mexico to Venezuela.

4-11: *La Caravana sponsors sustainability workshops wherever it travels, such as with this Permaculture Course in Colombia.*

Seven Workshops in Colombia

Where: Medellin, Darien Choco

When: September/October 2000

Who: The Reserva de Sarsardi (a 15-year-old ecovillage in the Darien jungle), the FUSDEUN (Fundacion Social para el Desarollo) of Unguia (a small outpost jungle town on the border of the Colombian conflict), and Montana Magica (an 18-year-old organic agriculture project recently converted to an ecovillage project). Instructors came from all parts of the Americas, and more than 200 students were trained.

What: Seven workshops in permaculture, natural building, ecovillage design, appropriate technology, and consensus decision making.

How: We received support from the Gaia Trust, the Ecovillage Network of the Americas,

the World Wildlife Foundation, and the Red de Reservas de Colombia. Fees for participants were reduced in many cases.

Three Workshops in Ecuador

Where: Two ecotourism sites in Puerto Quito, Ecuador

When: July/August 2001

Who: Fifty enthusiastic students participated, taught by US-based Becky Bee (author of
several well-known books on cob construction). In all over 100 students from Ecuador,
Colombia, Peru, and the US were trained.

What: How to build with cob (students created a cob oven and bench). Other workshops in the series included "Introduction to Permaculture" and "Consensus
Decision-Making." In July 2002 Becky Bee led a follow-up workshop. This successful
workshop has spawned groups dedicated to natural construction, permaculture design,
consultancy and education, and alternative economics. Javier Carrera (an ex-Caravanero)
has been instrumental in the development of these workshops and is now helping to
organize the Ecuadorian Seed Bank Network.

How: Participants' fees and a grant from the Ecovillage Network of the Americas funded
the workshops, and many students received partial or full scholarships.

Cultural Recuperation and Resistance in Agua Blanca

Where: Agua Blanca, Ecuador

When: January 2002

Who: The villagers of Agua Blanca, a remote indigenous community on the coast of
Ecuador.

What: Workshops to enhance the recuperation of ancestral roots. Initially the workshops focused on the arts, the making of stilts and drums, and music and dance. As
relationships deepened we began to introduce themes such as permaculture, appropriate technology, and natural building. (We had much to learn since people there have
been using adobe and bamboo construction for centuries.) Our month-long work cul-
minated in a community theatre presentation that reenacted the 5000-year history of
the village. Costumes, instruments, stilts, drums, puppets, songs, and dances were created for the event and visitors from other villages provided a most appreciative
audience. A video entitled "Recuerda Agua Blanca, Recuerda" (produced by La Caravana's
video team) documents the innovative techniques used in this project and will help the
villagers and La Caravana to further support the resistance of indigenous peoples to de-
culturization.

Women's Peace Village

Where: Azuay, Ecuador

When: June 2002

Who: Two women's groups and 140 women leaders from across the country

What: A three-month-long training of two women's groups. Diverse workshops included consensus decision making, organizational skills, event planning, self-esteem, permaculture design, and video documentation. With the women's groups we planned and executed the first weeklong Women's Peace Village, bringing women together to learn, share, celebrate, inspire, and be inspired. Diverse workshops (including permaculture), councils, plenaries, and multicultural sharing helped create unity amid diversity. A video entitled "Tejiendo Redes, Tejiendo Futuro" ("Weaving Networks, Weaving Future") was produced to share information and encourage replication of the process. A consensual document launched on October 12, 2002 formed the Ecuadorian Women's Leadership Network for Peace.

How: A grant from the US Department of State funded the initial training and the Women's Peace Village.

CHAPTER 5

Tell, Show, Do: Teacher Training Programs for Tomorrow's Housing Solutions

Melissa Malouf

Technology transfer is the act of introducing a new method or machine to a group previously unaware of it. This is a centuries-old occurrence with a newfound name. It is seen in events as simple and influential as introducing local material cross-bracing for hurricane safety, or as complex as introducing an entirely different roofing system to counter deforestation (see "Woodless Construction"). The intent of appropriate technology transfer is to find a technology that is in keeping with the particular goals and aspirations of a community and create an acceptable way to convey that information. To complement local tradition and society new technologies will need to be adapted for the culture.

There are many reasons to adopt a technology. For instance diminishing natural resources may necessitate the use of such resources in a new and more efficient manner or the introduction of new materials and the know-how to build with them. Natural disasters may lead to needed improvements of local techniques. A balance is found when "modern" components and improved technologies are implemented incrementally, while overall building structure or construction materials remain close to traditional methods. The optimum situation is when a new technology builds on traditional methods and materials and enriches the lives of those involved.

Cultural appropriateness is paramount in successful technology transfer and takes into account economics, ecology, tradition, society, and other relevant factors. An appropriate technology also makes the best use of the resources available. In "developing" countries, this is often human power, as labor is abundant and low cost. Thus programs and building systems should be labor intensive rather than capital intensive. The most appropriate materials are produced on a small scale, with little capital investment.

Furthermore all programs should concentrate on sustainability and practicality of long-term growth.

Disseminating the Information

Participatory Learning

One of the major disadvantages of the larger construction plans laid out by governments or agencies is the lack of participation by residents. Participatory training, in which the builders have hands-on experience, has proven to be an integral part of successful building improvement programs and has led to the adoption of new housing improvements. The participatory style is a learner-centered approach. The focus is on the participants developing their abilities and skills to diagnose and solve their own problems. The learner's needs, experiences, and goals are the focus of the training. The trainer merely facilitates a process of learning. Learning takes place more readily when trainees actively process information, problem solve, and repeat new practices.

Participatory programs foster teamwork and enable communities to work for the greater good of the population. Even when aid may cease, skills to build safely and economically remain. Participation is needed at all levels to assure that aims are agreed upon and met. The people themselves carry out this process and therefore it belongs to them.

5-1: Canadian architect Jean D'Aragon, teaching Brazilian youth the basics of building with compressed earth blocks.

Horizontal Learning

To maximize success, it is best in building construction to change the conventional Western teacher/student relationship. Indeed the formal classroom setting may not be a recognized form of learning in many societies. In "horizontal learning," knowledge is shared between trainers and trainees, with no subservient positions, allowing for a more open exchange of ideas. This allows the trainees to learn, experiment, and contribute their own knowledge in a safe environment. It also allows trainers to learn from the experience, skills, and goals of trainees. The passing of information from top-level authorities down to local builders creates a subordinate situation, which is not conducive to a true understanding of the subject.

The Importance of Mobile Training

The widespread dissemination of information is essential to the large-scale appropriation of housing improvements. By bringing the information to the people, new technologies can begin to infiltrate the rural built environment. Mobile training allows for adaptations to specific sites so that improvements will be appropriate for each location. This is especially important if conditions vary widely throughout a region.

Without incentives it is unlikely that people in remote villages will have the time or resources to attend special training courses in distant locations, and if so, transportation is often unavailable. If builders need to continue working for wages, it is very difficult for them to attend unpaid training. To be most effective training should take place on-site, where people are already active in rebuilding their own homes and can allow for a flexible schedule.

The Use of Educational and Training Aids

Educational aids and training materials often include illustrated manuals, flip charts, posters, videos, and presentations. Some of the most successful building improvement projects, however, have been conducted without the use of educational aids and manuals. Nevertheless, when used appropriately, training aids can help in the comprehension of new ideas. These materials can stress main points that have been covered in training, and many projects produce such materials to transmit new information. However all evidence points to the fact that training materials have little or no impact when the rest of the necessary supports for a new technological innovation, such as skills training and building materials familiarity, are lacking.

Understanding cultural constraints is vital to any education program, and lack of understanding of a specific cultural situation or inappropriate content or format can mean that visual aids become totally ineffective as a means of conveying the intended information. Printed material can be improved through integration of field experience; in fact training materials and educational aids should be seen more as a final product of the process rather than as a necessary tool for the initiation of a project.

In no case should educational materials take the place of a trained educator. A skilled advisor who can successfully explain a concept can transcend poorly made educational materials. However even the most beautifully drawn graphics still need commentary.

DEVELOPMENT WORKSHOP

5-2: *Development Workshop uses special training structures to educate students about the basics of woodless construction. In this example in Mali, students learn how to build vaulted roofs of adobe bricks.*

Training Methods

Variety

Basic methods of training can be briefly described in three words: "tell," "show," and "do." The first form, "tell," is regarded as the least effective of the three. It requires a large number of materials and a high degree of motivation and commitment on the part of the trainer, as well as listening ability on the part of the trainees. The academic style is a content-focused educational approach. Information is primarily passed in one direction, from outside expert to student. Learning through a class lecture and taking notes is usually a skill acquired through years of formal education. It is not reliable as a productive form of teaching where people lack this formal education.

"Show" can take a variety of forms and increases learning capabilities. Visual aids and site visits help trainees retain information. Still, in the "tell" and "show" approaches, an unmotivated trainee may appear to be listening but, in fact, may not be fully participating and retaining the information. The "do" methods — practical hands-on building or interacting with a model — involve the participants in a physical activity and make it easier to learn.

A combination of educational methods is most effective when teaching new and modified building techniques. A series of courses should be given, each one concentrating on one or two main ideas, with adjustments made as the course progresses. Depending upon the topic, one class may be enough to impress the main idea. At other times, an entire sequence of presentations may be necessary. Actual building experience must be integrated into the program. However it is also helpful to have a scale model for examining detail. In these ways, concepts are reinforced with the use of various training methods and materials. To convey critical construction details the practical, hands-on aspect of a training course should be extensive and involve repetition of several tasks over a period of time.

5-3: Argentinian students display a model house created during an ecological design and natural building course held at the Associación Gaia Ecovillage Project.

Guidelines for Educational Aids

Illustration

Illustration is an effective teaching aid but should be tested for comprehension within the community. When images are clearly understood they are a good way of contrasting safe and unsafe practices. In Dhamar, Yemen, faced with a high illiteracy rate, Oxfam found that "P" and "X" marks were understood as correct and incorrect (see Text Box: The

Dhamar Building Education Project). However in many other cultures, pointing hands and incorrect marks only led to misunderstandings. Take care to draw realistically and avoid exaggerated sizes and abstraction. Perspectives and true-to-life images aid in comprehension. Avoid unneeded detail and emphasize the most important points.

Posters

Posters should be used only to convey general information, (i.e. time of meetings, program goals, and general concepts), never in order to illustrate technical points to the public in place of training. Without explanation, posters cannot convey critical detail or ensure that important points are understood.

Video

Oxfam has stated in its evaluation of The Yemen Building Project that they feel the value of video films as a training tool is limited, as the equipment might be difficult to obtain. Furthermore, in some regions, use of the video and television may prove to be the focus, thus distracting from the information itself. While video has potential as an awareness-raising tool, it is a major undertaking for a project. The best way to demonstrate a tangible point is to enact it in practice. However video is highly successful in showing damage to unsafe buildings and the potential of new innovations.

5.4: *At the Bethel Business and Community Development Center in Lesotho, students learn by constructing the actual buildings of the campus from foundation to roof.*

Models

Models can be vital to the comprehension of construction details that are difficult to convey through drawings. Scale models of buildings should be kept very basic, illustrating only one point per model if possible. Models of the complete dwelling may sometimes be needed if builders are having difficulty conceptualizing a component of a building without the context of the whole.

Use of Training Structures versus Actual On-site Training

Some programs use training structures to hone building skills; others train builders through constructing actual dwellings. Programs should incorporate whichever option seems most fitting, as choice is sometimes dictated by situation.

If using training structures, the idea that the building is "only an exercise" can inhibit true comprehension of the technique. Hands-on training on actual construction sites compels trainees to build safely and correctly, knowing there are consequences to improper construction. On the other hand, during on-site learning situations, the deadline pressures that go along with a real building project can result in uncorrected errors that detract from the learning process and possibly foster future inaccuracies. Training structures allow for the repetition needed for the acquisition of new skills: when there is doubt about a technical detail, time can be allocated to review the information, assuring comprehension.

Optimally both methods would be used, with training structures used for repetition of difficult components, new safety features, and details. The subsequent construction of a demonstration building would reinforce the newly acquired information, while applying it in practical, real life circumstances.

5-5: On its travels through Latin America, La Caravana uses simple posters to educate often illiterate villagers about different aspects of sustainability.

Building in Post-Disaster Situations

Post-disaster situations dictate whether or not demonstration buildings can and will be built. If lack of shelter is a pressing issue and time is of the essence, construction may need to take place immediately as a "learn as you go" program. In such situations supervision is paramount, and it is crucial to understand the established norm of "transmission of knowledge" and apprenticeship. Disaster situations accelerate the learning process and can cause problems for students that may lead to construction errors.

To pass on new knowledge, one must understand and respect the community and which approaches would work best with the people involved. Through explaining what improvements are proposed and how they will contribute to safe, improved dwellings that are economically attainable, climatically effective, and sustainable, housing can be made most effectively available to those that need it.

IMPROVED QUINCHA, ALTO MAYO, PERU

Intermediate Technology Development Group (ITDG)

In May 1990 an earthquake measuring 5.8 on the Richter scale hit the Alto Mayo region of Peru in the northeastern Amazon, destroying over 3,000 houses. In April 1991 another earthquake, this time measuring 6.2, damaged and destroyed another 9,663 houses. The rammed earth (*Tapial*) houses predominant in the region suffered the worst damage, with 80 percent of buildings damaged. However *Quincha* building (wood cane and mud infill), introduced from neighboring regions, proved to be earthquake resistant. Six months after the first disaster, *Tecnologia Intermedia* (IT Peru, a branch of ITDG) initiated the Alto Mayo Reconstruction Plan. IT Peru introduced modifications to traditional building technology to reduce vulnerability and improve durability.

ITDG proposed using existing knowledge to improve local building practices, focusing on Quincha, which only required minor improvements to increase durability. Through the reconstruction of buildings, the dissemination of appropriate building techniques, and a reforestation awareness effort, ITDG helped people adopt a safe building technology to rebuild their homes. The program emphasized use of local resources such as labor, materials, organization, and management. Specific structural changes included concrete foundations to aid in stability, integrated wooden structures that attached foundation to roof, and the use of plaster to guard against humidity and to increase social status and acceptance.

Apart from earthquake resistance, a key aspect of the technology was its adaptability, allowing it to reflect local conditions and the aspirations of the residents. Adapting Quincha technologies involved extensive discussion with local people about their housing needs. The improvements ensured that it offered a modern appearance, vital in ensuring a widespread acceptability in the region. Low cost and simplicity in design allowed for a high degree of self-reliance.

Project beneficiary families were taught basic notions of architectural design. Using colored bricks to identify each space within the home, participants created simplified plans of their future homes. ITDG then trained local builders, using on-site technical assistance during construction of several demonstration buildings. All builder training took place on the job. ITDG developed their training activities specifically for local construction masters, with the goal of creating "promotion agents" for the ▶

improved Quincha technique. However many builders were unable to stay for the entire process because they had to work.

No formal training courses, written, or audiovisual materials were used. As of 1995 no instruction manual or training materials had yet been produced on the improved Quincha technology. The lack of a training manual demonstrates that communication between promoter and user is far more important than any amount of training material. No attempt was made to introduce a pre-designed technology package. The improved Quincha system built directly on local skill and design. These factors ensured the long-term adoption of the improved Quincha technology. The 70 houses constructed directly after the first earthquake were a deciding factor in the success of the program. These houses suffered no damage in the second earthquake, demonstrating the innovation's legitimacy, which then led to acceptance.

The Peruvian government has incorporated the improved Quincha into its own projects, showing that the program has taken hold in a larger context. This government promotion encourages the public to implement the technology. Due to the wide adoption of the building innovation by local families, this program is considered a success.[1]

THE DHAMAR BUILDING EDUCATION PROJECT, YEMEN

Oxfam, Redd Barna, and Concern
In December of 1982 an earthquake in the Dhamar region of Yemen left approximately 300,000 people homeless. Typical traditional homes in Yemen are built of massive stone masonry with rubble and mud mortar. They are very vulnerable to earthquakes, due to lack of reinforcement and weak construction details. To remedy this situation, the Dhamar Building Education Project (DBEP) proposed to promote awareness of the need for safe building among the general population and the teaching of simple, improved building techniques to builders.

This project was not directly involved in the reconstruction of buildings. The aim was to promote a set of simple technical messages geared towards local builders to encourage the construction of earthquake-resistant dwellings. The project advocated improvements in foundations, the incorporation of horizontal reinforcement in walls (including ring/tie beams), and the repositioning of windows that would have ▶

contributed to weak corners. These changes were incorporated while staying true to traditional forms.

DBEP representatives worked closely with local professionals to develop materials that would be appropriate for the local population. Local resources and traditional construction methods were used as the basis for any improvements. Training consisted of a three-day course, covering the principles and techniques of strengthening buildings in a practical manner. Participants built a training structure that detailed all aspects of the improved construction. The course used a short manual that stated main points, though this illustrated manual was never intended to be self-explanatory. The manual was explained, page-by-page, and updated as needed. Builders were required to pass a verbal examination before being awarded certificates.

Experienced local masons conducted the courses after being trained by the core organization. A training center in Dhamar provided ongoing information and technical support, while DBEP employed a Mobile Education Unit (MEU) to reach the surrounding regions and spread the information to as many people as possible. Videos played an important part in the training methods of the MEU but were used as aids, not as the main source of training.

Flexible training allowed for a response to individual problems as they arose. Changes in structure or material were only made after discussion with the local communities to assure that all innovations would fit socially and culturally. As storytelling has a strong history in Yemen, oral presentation of techniques helped disseminate the information; discussion and dialog were more important than visual aids. From this experience the DBEP believes that the spoken word is the most effective way to communicate with a population that is not formally educated.

The DBEP and MEU encouraged people to build houses that are safer and that preserve traditional Yemeni architecture. Toward this end they attempted to reinstate the forgotten tradition of tying beam ends into walls. Omitting this practice contributed significantly to the collapse of roofs during earthquakes. By reinstating this simple reinforcement technique, Yemeni builders could build earthquake-resistant structures that also had historic character.

The training project visited 250 villages and trained over 800 villagers. A post-project review revealed a good knowledge of safety amongst builders several years after the initial training. Adoption by the wider community was slow, however, due to an economic recession that reduced the opportunity to build new houses.[2]

Case Study:
Teaching Sustainable Settlement Design in Lesotho

Ivan Yaholnitsky

Traditional Settlement in Lesotho

Prior to European influence, Lesotho was primarily a summer grazing area, and the Basotho people migrated to lower altitudes for the winter, where they constructed ephemeral beehive-type structures from woven reeds, grass, and poles. The vagaries of geopolitics, however, disrupted this pattern, and following the Anglo-Boer wars, seasonal migration ended. Thus began the contemporary era of fixed settlements and more durable housing.

Traditional settlements are located on north-facing, mid-slopes of hills and valleys, where micro-climatic conditions are moderate. These thermal belts are preferred for settlement in virtually all cases. Houses (rondavels) are constructed with walls of thick brush and mud or stone and mud, which serve to balance diurnal heat gains and losses. Doors face north to allow winter sunlight inside. Thatched roofs insulate by retaining winter heat and providing good shielding from summer sunshine. Biomass fires augment heating.

5-6: *A traditional rondavel.*

The rondavels are often set partway into an earth slope and back-filled with an earth berm, further serving to moderate interior temperatures. The circular design of the rondavel is extremely stable because of the equalization of stresses. Taken together, this combination of positioning, orientation, structure, and materials is a good passive design that tends to be warm in winter and cool in summer.

Drawbacks of this type of housing include limitations on size, darkness because of a lack of windows or artificial lighting, dampness and migration of moisture through the floor in wet weather, and poor indoor air quality because of a lack of ventilation. Despite these shortcomings, there is much to commend the patterns of settlement and shelter developed by the Basotho.

Contemporary Trends and Directions

Modernization in Lesotho has not achieved the expected benefits. As new materials and methods of construction have been adopted, former patterns of livelihood, common sense, and settlement have been abandoned. Urbanization and the sprawl of settlements along the lowland corridors between the major cities have accelerated over the past 50 years. Traditional housing styles have been largely abandoned in favor of those of concrete or brick with corrugated iron roofs.

If sited and oriented badly, such "modern" buildings can be very inhospitable and far inferior to traditional thatched rondavels. During summer the corrugated iron can reach temperatures over 175 degrees Fahrenheit (80 degrees Celsius), making interior conditions unbearable. In winter the corrugated iron cools rapidly after dark, and if indoor cooking and people's breathing has added humidity to the indoor air, condensation drips throughout the night. Steel window frames and single panes of glass contribute further to rapid heat loss in winter. To compensate, biomass, coal, or paraffin fires are used extensively, which leads to dangerous indoor air quality.

For most of the population, the overall picture of housing in Lesotho today, then, is less a story of comfort than one of endurance, suffering, excessive energy costs, environmental degradation, and toxic indoor air quality. Even in cases where households have access to relatively clean electrical power, it remains expensive.

Under pressures of resource scarcity there has been a rediscovery and corresponding appreciation of indigenous practices, along with a reevaluation and criticism of many modern technologies.

Bethel Business and Community Development Center

The Bethel Business and Community Development Center (BBCDC) was formed to tackle some of the most challenging settlement problems facing Lesotho. BBCDC is a school for rural and urban development, located in a remote region of Lesotho. Since being founded in 1993, its main goal has been to construct low-cost, innovative, and vernacular-styled buildings through self-help and authentic education. To date, staff and students have constructed 15 major stone masonry structures on the BBCDC campus, including work buildings, staff housing, student dormitories, and guest facilities. BBCDC has built another 7 structures off-campus. All buildings were constructed as a "learn by doing" exercise. A member of local development groups, BBCDC hosts a steadily increasing stream of study tours and visitors from throughout the region.

The campus is a living laboratory, with every building pushing the envelope and attempting something new. Currently 50 men and women attend a two-year, holistic cur-

5-7: *Simple metal slip forms allow for accurate stone walls to be built relatively quickly. This system is based on the work of sustainable building pioneer Ken Kern.*

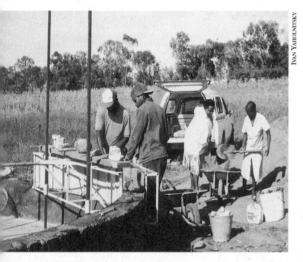

IVAN YAHOLNITSKY

5-8: Students, building the stone walls of one of the campus structures in Lesotho.

riculum with hands-on exploration of subjects in solar technology, agro-ecology, building construction, metal work, woodwork, home economics, business, and environmental education. Students learn by problem solving on real world, practical projects. Students work on all facets of building projects: foundation, floor, walls, roofing, finishing, plumbing, electrical wiring, the building of furniture, various solar technology systems, and landscaping. They are encouraged to draw as much as possible in order to develop design skills, and woodwork and metalwork develop support skills for construction work. They also spend time on mapping and the use of various survey tools and equipment. We are developing our own style and thoroughly incorporate water harvesting and solar energy strategies into our structures, as well.

Focus on Authentic Education

Our belief is that sustainability is not just a moral issue but also one requiring wide-ranging technical and social competence, combined with innovative educational processes.

The main objectives of education at BBCDC are:

- To instill a spirit of dedication and teamwork in the pursuit of common objectives
- To equip students with practical skills related to the needs of rural development
- To develop an understanding of appropriate technology and natural resource management
- To develop capacities for building household and community infrastructure through use of basic techniques and local materials
- To develop resourcefulness and problem-solving skills
- To analyze those forms of technology and social organization that foster harmony between people and the environment.

The course places special emphasis on:

- General engineering and basic skills
- Collaborative processes
- Solar energy utilization
- Water resources development

- Environmental regeneration and household food security
- Rural and urban infrastructure
- Village and peri-urban design
- Appropriate technology.

Theoretical lessons are brief, with most time spent on self-help or income-generating projects. The holistic nature of the curriculum and the two years of work and study at the school become a story of accomplishment. The mix of activities elicits more than cerebral learning: real work situations develop visual perspective, motor coordination, and social skills. These faculties cannot be taught abstractly but must be practiced and experienced. Since the need for infrastructure is so high, each student can count several major projects in which they have participated, from groundbreaking to completion (no outside contractors or day laborers are used at the school for any of the infrastructure.) We are greatly oversubscribed for new students, and in 2002 the Department of Manpower made a decision to pay all the students' fees at BBCDC. We did a follow-up study of graduates in 2000, and 75 percent had work or were fully self-employed.

IVAN YAHOLNITSKY

5-9: A view of the adult education and guest facility. Built in 1996-1997, this building totals 1400 square feet (140 square meters) of floor area and consists of a large central classroom, four bedrooms that sleep eight to ten guests on two wings, and two bathrooms. Both walls and ceilings are insulated with fiber-glass wool, and additional features include daylighting, a roof-integrated solar water heating system, and 2-by-75-watt photovoltaic system for lighting.

Strategies for Sustainable Construction

Our integrated building design process includes the following concepts:

- Vernacular style, local materials, and simple construction processes
- An integrated set of solar energy systems for daylighting, solar water heating, space heating, ventilation, and electricity
- Water harvesting
- Intensive use of terraces, containers, and trellis-wrapped buildings for use of roof water for food production
- Landscaping for temperature moderation
- The recycling of domestic wastes, grey and black water.

BBCDC uses a "shuttered stone masonry" system adapted from the work of Ken Kern. Following close analysis of Kern's work on the subject, BBCDC opted for a fairly large modular system for building shutters. These are built at a height of 17 inches (44 centimeters) to achieve standard wall height in six consecutive lifts. Forms are built in straight configuration from 80 to 110 inches long (200 to 280 centimeters long) (in pairs) and in circular arcs with a radius of 60 and 110 inches (150 and 280 centimeters). Straight and curved forms can be combined to produce imaginative floor plans.

Although the shutters make building accurate stone walls much simpler and faster, they are not a substitute for a careful stone-laying sequence, attention to detail, and finishing. The best appearance is obtained if stones are packed as closely together as possible, with staggered vertical joints and "tie-backs" of long stretcher stones from windows and door openings. Runny concrete that smears rock faces must be avoided. Neat finishing of recessed external joints and stiff mortar produces a professional look-ing wall, if combined with the cleanup of cement stains with a weak hydrochloric acid solution.

Reinforced-concrete lintel beams over windows and door openings are cast in situ. Wire masonry reinforcement is run continuously through the walls, and tripled on the last course to form a tie beam. Wire reinforcement cast into the wall is also used to tie down the roof. (Windstorms are notorious in Lesotho for removing roofs from con-ventional buildings.)

In March 2002 BBCDC completed the walls for a shuttered stone masonry build-ing, consisting of a new kitchen and classroom. Innovations include daylighting, a roof-integrated solar water heater, rainwater harvesting, a solar chimney for ventilation, and an advanced photovoltaic (PV) power system that will not use any DC wiring. Compact AC fluorescent lights have just become available in Southern Africa, which means that PV systems can now be designed from the beginning to only operate AC loads. This is more efficient and safer.

Strategies for Water Harvesting

In arid environments water is a precious resource. Without careful bioengineering, energy flows through a site wildly with wind, rain, sunlight, and temperature conspiring to torment the inhabitants of the built environment. Intelligent landscape design, how-ever, is capable of rendering the sun soothing and mild, the rain restorative and cleansing, the wind refreshing and cool, and general climate temperate.

Large impermeable surfaces such as roofs, paths, roads, play areas, and parking lots must be designed properly to prevent torrential flooding and erosion and to conserve water. BBCDC designs allow for the incremental collection of water that is put to good use elsewhere. In many instances, roof water provides for all domestic consumption and use with enough surplus to create surrounding food gardens. And this water conservation technique alone can largely eliminate the need for supplementary cooling of small buildings.

Conclusion

BBCDC is searching for models of economic growth and prosperity that do not jeopardize our children's future. Growth does not have to mean an irreconcilable trade-off between the environment and the economy. Environmental rehabilitation, renewable energy, integrated building design, landscape design, and new patterns of livelihood, community, and education can be the engine for growth that is sustainable both in the short- and long-term.

WHY WE DON'T ROOF WITH THATCH

People always ask us why we do not roof our buildings with thatch. Here are the reasons:

- Our local environment is so badly degraded that we would have to go nearly 2,500 miles (400 kilometers) to find thatch — this is not cheap.
- Southern Africa lags badly on many environmental fronts: poles are still treated with creosote, making them unhealthy and a fire hazard.
- The cost of a pole/thatch roof is anywhere from four to six times the cost of a metal/frame roof.
- Thatch does not lend itself easily to daylighting, the fitting of PV panels, solar water heaters, or the collection of rainwater. In addition the roof rats here are the size of badgers and can eventually bore through thatch. I have also seen the results of several thatch fires and, for this reason alone, would not house my family in a thatched house. (Southern Africa suffers extremely gusty spring winds, and when thatch burns, it burns fast.)

All told, I like corrugated iron; it is cheap, fast, and allows for the easy collection of water — a most precious resource in these parts. Moreover when you have large institutional buildings to put up on a tight budget, it is the most economical option.

Case Study:
Capacity Building on the Periphery of Brasilia, Brazil

Rosa Amelia Flores Fernandez and Jean D'Aragon

Brasilia, the capital of Brazil, has been on the UNESCO World Heritage List since 1987 and is well known by architects all over the world. However Brasilia also stands out as the most explicit example of "Brasilianization" — a phenomenon whereby the poor are continuously attracted to the city yet forced to live in peripheral settlements. Satellite cities beyond the city limits *(Entorno)* contain 75 percent of the population of the *Distrito Federal* (Capital State).

5-10: *A typical shanty in Santo Antonio do Descorberto on the outskirts of Brasilia. This is the prevalent housing type for this city.*

Situated 22 miles (40 kilometers) from Brasilia, the city of Santo Antonio do Descorberto is typical of the periphery of Brasilia in that it has seen tremendous recent growth — its population increased from 60,000 in 1997 to 100,000 in 2001. During that time, Santo Antonio was unable to provide even minimal infrastructure (shelter, water, electricity) for the people who live there. The population is predominantly young people, most of whom are dealing with issues relating to drugs, gangsterism, or prostitution. Families average seven to nine children, and children start to work as soon as they can in order to contribute to household income. Overcrowded living conditions often lead to incest.

The landscape is rough with many ravines, valleys, streams, and rivers. The region has well-defined dry and rainy seasons, with the rainy season extending from October to January (responsible for 80 percent of the 67 inches (170 centimeters) of the overall annual rainfall). This brief but intense period of precipitation causes many landslides and floods, and soil erosion is high in this heavily de-vegetated area. People settle in ecologically fragile zones, 60 percent of which are prone to landslides. The largely uneducated population lives in precarious conditions, dwelling in shacks on dangerously steep slopes. Continued growth amplifies environmental and health problems, as people are both responsible for and victims of soil erosion, landslides, and pollution. As households are frequently established near streams, they add to and/or are in contact with the existing pollution coming from upstream. According to the 1999 census, up to 10 families (at least 50 people) were living on 3,875 square-foot (360-square-meter) plots, frequently in hazardous conditions. Beside the current housing backlog of 5,000 housing units, 2,000 new units are needed every year in order to meet the current population growth. Of this needy population, only 35 percent will find a shelter through the formal market. The 65 percent that cannot afford to buy a house dwell in shacks called *barracos* built as squatter settlements or in rented backyards.

Permanent dwellings are most commonly built from burnt clay bricks and concrete blocks, both of which are economically and ecologically costly. All cement, concrete blocks, and the very popular extruded bricks are produced outside Santo Antonio. Besides the high embodied-energy and transportation costs, the production of these materials generates little if any economic gain for the inhabitants of Santo Antonio.

Solid clay bricks are used for foundations and septic tanks, but rarely for walls because of their uneven quality and low strength. The bricks are produced by a few small-scale brick factories located along a local river, but the few positive impacts on the local economy, and the limited transportation costs of this low-tech product are negated by their environmental impacts. Factories extract the clay directly from riverbanks, worsening the erosion caused by deforestation, which itself results from the harvest of firewood for the inefficient and polluting brick kilns.

Houses are often built in the course of a few days. Qualified bricklayers are largely unavailable outside Brasilia proper, and households build their dwellings with the help of unskilled workers. These workers, largely unaware of good building practices, often skimp on materials such as reinforcing steel or cement. They may be too generous with water in the mortar mix or prepare it too much in advance, leading to greatly decreased building strength. The resulting shoddy, yet expensive structures are often the first to collapse during the rainy season.

5-11: *Students mix clay, sand, and water in proper proportions to make compressed earth blocks.*

Sustainable Development for Santo Antonio do Descuberto

Soon after the municipal elections of 1996, the mayor of the new administration of San Antonio invited architect and urban planner Rosa Fernandez (co-author of this article) to create a Department of Architecture and Urbanism. One of her first tasks was to perform an environmental diagnostic and establish a master plan for the city. The innovative plan was the first in the periphery of Brasilia to integrate a system of environmental protection.

The Department of Architecture and Urbanism also developed some low-cost housing projects, by taking advantage of different Brazilian government and cooperation programs. To address the lack and quality of housing, the projects created an "evolutionary house" concept, using assisted self-help construction methods and the help of a master mason paid by the municipality. Cooperative organizations developed credit schemes with low interest rates to allow the poorest to access ownership. Between January 1998 and December 2000, these initiatives resulted in the building of 182 housing units (for the most part starter units that could be expanded over time).

Encouraged by the positive results, Rosa saw the planning of Santo Antonio as part of larger effort of "sustainable development. With her architectural background, Rosa understood that the choice of a material or technology is never without social, political, or ecological impact. She set her heart on using housing as a lever for local development and, at the same time, reducing its negative impacts on the environment.

Earth As a Sustainable Material

Rosa was familiar with the durability of earth buildings. A native of Peru and a specialist in earthen building history, conservation and restoration, Rosa knew that earth still shelters more than 60 percent of the Peruvian population. She also knew that earth could be part of the solution to the housing, environmental, and unemployment problems of Santo Antonio. So in June 1998 she attended an intensive "Low-Cost Housing and Earth Construction" course at CRATerre (International Centre for Earth Construction) in Grenoble, France.

Jean D'Aragon was a researcher in residence at CRATerre at that time, and our meeting led to the creation of a training program in earth technology, as well as to a low-cost housing project to address Santo Antonio's needs and aspirations. Aware of the importance of starting slowly, our idea was to set up a project that would demonstrate the feasibility of our approach and then grow to address ongoing housing needs. Our search for international funding was unsuccessful, but the project proceeded and eventually generated some unforeseen outcomes.

5-12: Students, using a hand-operated press to produce compressed earth blocks. The students formed a cooperative to produce this sustainable resource, with the assistance of Jean D'Aragon and Rosa Fernandez.

Teenagers As the Pivot for Sustainable Development

Rosa returned to Brazil to pursue her work as head of the Department of Architecture and Planning in Santo Antonio. At this time, the municipality was selected by the Brazilian government to initiate a program to help teenagers from poor families prepare for the future. Twenty-five teenagers between the ages of 15 and 17 were selected to participate. Some were afflicted by family and social problems such as alcohol and drug consumption, stealing, or prostitution. Participants received an allowance to prevent them from quitting the program and/or school to go to work. And the Social Assistance Office offered psychology, communication, sociology citizenship, and leadership courses to boost the teenagers' self-esteem as future agents of change and development in their communities.

Although the federal program was about preparing the teenagers for the working world, we took advantage of the situation train them for a new market — sustainable development. We wanted to teach them how to produce building components that would use local resources having fewer negative impacts on the environment and an improved impact on the local economy.

As traditional techniques such as adobe and wattle and daub were not acceptable to the urban dwellers of Santo Antonio, we finally selected Compressed Earth Blocks (CEB) as our technology of choice. (CEBs were a good, inexpensive alternative to the typical building materials used in Santo Antonio.) Material to make the blocks (70 percent sand and 30 percent clay) could easily be found locally, and the blocks themselves did not need to be fired. When produced under good conditions, the blocks could be stabilized with minimal cement content. The relatively low-cost material relies on local material and human resources, and has a limited impact on the environment. And because of the "soft" industrial process used to manufacture the blocks (manually operated presses) and their appearance (similar to industrialized products), CEBs have been well accepted by both producers and final users.

5-13: *In order to encourage a market for compressed earth blocks, Jean D'Aragon, built a prototype building of the new material in downtown Santo Antonio, with the help of a local mason.*

Having been invited by the University of Brasilia to give two courses on straw bale and earth construction at its Faculty of Architecture and Urban Planning, Jean took this opportunity to cover his passage and stay in Brazil so that he could join Rosa to work on the training program. Although we had no block press or the means to buy one, we nevertheless started training the teenagers, working with the materials and tools at hand. We realized that we should not be too purist in our approach. Our experience told us that a series of small-scale projects with humble results are better than ambitious projects with no results at all.

Knowing that the teenagers would eventually own their building material cooperative and that their activity would have to be profitable, we started with the production of a material that already had a good local market. We salvaged an abandoned vibrating machine from the municipality's works yard and trained the teenagers to produce concrete pavers. The pavers were a desirable commodity, compared to conventional concrete slab paving, as they reduced the use of concrete (because of their minimal thickness) but were very strong (because of the vibrating process and controlled curing). As the teenagers learned about paver production, they were simultaneously learning about different soils and aggregates, stabilization, dry and wet materials mixing, as well as the curing process of cement-stabilized building components — all important aspects of CEB production. Several weeks later when the teenagers had gone to and returned from a course on starting and managing a cooperative business, we finally found two CEB presses and could continue with training.

Our efforts generated interest in several cooperatives and other organizations and led to the creation of the first Building Materials Cooperative held by teenagers in Brazil. Concurrently, with the help of an old mason to whom he introduced CEB masonry techniques, Jean designed and built a canteen at the public market as a way to open the market to the young "co-operators" for their new building material. This was an inexpensive and effective way to disseminate the technology and build a demonstration model in the heart of the city.

We will see if the seed we planted will survive and grow. We faced a setback when, in the next election, the supportive incumbent administration suffered a defeat by a more conservative group. And we now realize that we were perhaps a little ahead of our time for the reality of Santo Antonio. But the young cooperative is still producing the building materials and providing other services to the community. And the cooperative movement, which is very strong in Brazil, continues to support the teenagers, who have become an example of what can be achieved with meager means if people unite and the political will is there. It is not always necessary to wait for aid from big international organizations to implement community projects. The will and energy of a few individuals can suffice when people gather together with a common goal and a firm commitment to the empowerment of the community.

We keep in mind the day the teenagers received their diplomas for their nine months of effort — the same day that the cooperative was officially launched, with members of the cooperative movement from all over Brazil assisting with the event. One could easily feel the pride that the kids had in their achievement and the impact that the show of support had on their parents and community. Being witnesses to that day was the best reward we could have had for our work.

PLASTERED STRAW BALE CONSTRUCTION IN CHINA

Introduction

Kelly Lerner

The Adventist Development and Relief Agency (ADRA) first introduced straw bale construction to Northeastern China in 1998 with the construction of a single school, but use of the technique has grown steadily since then. As of 2002 there were over 600 passive solar straw bale buildings in the northeastern provinces of Heilongjiang, Liaoning, Jilin, Hebei, and Inner Mongolia; most are houses, ranging in size from 540 to 860 square feet (50 to 80 square meters).

ADRA works at the national level with the Center for Sustainably Sound Technology Transfer (CESTT), a UN Agenda 21 project and part of the Ministry of Science. CESTT introduces ADRA to local communities who have both a need for housing and an interest in sustainably sound building techniques. ADRA partners with these local governments on both existing and new housing projects. ADRA also provides management and technical training and subsidizes each new house as an incentive to learn new techniques (average US$750/house; average total house cost US$2,375). The remaining house cost is divided between the local government and the new owners. ADRA also provides training to the new owners on maintenance and ecological impacts. Since the initiation of the project, many people have come to ADRA (on their own) seeking assistance to introduce straw bale building technologies in their areas.

Housing has contributed greatly to deforestation and the depletion of other natural resources in China. Additional ecological and social impacts include desertification in Inner Mongolia, air pollution in Northeastern China, and the relocation of people out of marginalized areas or flood plains.

Traditional housing patterns vary by location but are typified by single family and extended family households, where older parents live with the oldest son's family. Owner-builders initiate building projects (usually a new house for their single son so that he can get married) and hire skilled workers to complete work.

KELLY LERNER

5-14: *Creating a brick and straw bale house in Benxi, China.*

Conventional new construction typically features high rock foundations, load-bearing brick walls, wooden roof structure (beam and purlin or trusses), and tile roofs. Older houses typically consist of low rock foundations with rubble and earth walls, wood roof structure (beam and purlin or rafters and purlin), and a tile or mud roof. Houses are generally built in the summer, when people are not busy planting or harvesting, and are financed out of pocket and with loans from extended family.

ADRA's success in China is directly related to their intensive training program and successful relationships with local partners (see "Down to Earth Technology Transfer" and "Seismic Solutions for Straw Bale Con-struction"). All buildings are designed by local builders or architects who have been mentored by a foreign special-ist working with the program from the very beginning. Initially I trained builders and managers, but as the ADRA staff gained experience, they took on management train-ing. As the project grew American writer and builder Paul Lacinski joined the project to assist with training and site visits. ADRA has focused its efforts on developing local talent and within the next five years, it anticipates that straw bale building systems will be fully transferred to local experts, with no foreign inputs required.

The Uses and Limits of Plaster Samples
Paul Lacinski

In China most new construction is of brick, and the plaster over it looks great. When this cementitious plaster has been applied over straw bales, however, it has tended to crack. In the interest of developing locally sourced plasters that work over bales, Kelly Lerner and I have produced many samples based on lime, clay, cement, (or mix-tures of those ingredients), both with and without pozzolans (pozzolans are materials added to lime plaster to speed the set, increase strength and hardness, or increase the thickness at which the plaster can be applied).

The samples have proven very useful, both for introducing new ideas to local masons and for exploring the properties of the available materials. Over time, how-ever, we have learned that samples have their limitations: they cannot address the wide range of factors that ultimately determine the quality of a finished plaster. These factors fall into three general categories: materials and process issues, application techniques, and cultural issues.

Materials and Process Issues

Plaster samples are usually small, which means that they tend not to crack, regardless of what their behavior would be on an actual wall. At least in China, it is very difficult to find a section of wall on which to do full-sized samples, because houses are built very quickly (typically in about a month).

Samples also need to be considered from a builder's point of view. Let's say someone comes along, prattles on a bit about vapor permeability and cracking, and leaves you with some 5-by-9-inch (13-by-23-centimeter) samples. What do you do? You beat on them with a hammer (or your fist) to see which one is most durable. Of course the sample with the highest percentage of cement always performs best in this test, because the test is designed to favor rigidity. Subtler considerations such as flexibility and moisture transmission, unfortunately, are difficult to measure in the field. Although the better builders are often intrigued by new materials that seem solid as samples, they can be reluctant to try them on the scale of a house.

The materials themselves also present limitations. Some of our best lime tests were with crushed brick as a pozzolan. Brick rubble is heaped all over China and is free for the taking, so using it seemed like a perfect idea. Unfortunately we never could come up with an inexpensive, site-scale crushing method. (The available agricultural hammer mills are underpowered.) For the samples I used a sledgehammer. But even in China labor isn't cheap enough to justify this method, and brick dust has yet to make its way into a straw bale plaster.

We also made several rounds of fly ash samples. The limitation with fly ash is that if the source coal has too high a sulfur content, some or all of the lime will be converted to gypsum (calcium sulfate, rather than calcium carbonate). This material is not suitable for exteriors, as gypsum is soluble in water. And as the coal varies from region to region, it is necessary to test the samples at each site. That isn't complicated, but once again it takes time — the samples must set for two or three weeks and then be soaked in a bucket of water to see whether or not they hold up.

5-15: *Brick dust is an excellent additive to lime for increasing its hardness. Here Paul Lacinski crushes bricks with a sledgehammer to produce the dust. This method is impractical for large amounts, however, and an alternative method has yet to be devised.*

Application Issues

The difference between a durable plaster and a weak one often has less to do with materials than with application techniques. A sample that has cured slowly in the shade, for instance, will not predict the outcome of the same materials troweled onto a wall facing the hot sun and left to dry out.

Our experience in China has been that application techniques are more of a challenge than materials selection. Chinese masons (who are highly skilled in getting material onto the wall and creating a hard, flat surface) are accustomed to plastering over pre-wetted bricks. The wet bricks do an excellent job of releasing moisture into the plaster during the cure, and thus people are not accustomed to spraying the wall to keep the plaster moist. I have seen recently finished walls where the bale infill panels show clearly as large dry rectangles, framed in a border of damp plaster that covers the brick structure. Not surprisingly the bale panels then crack. Such a situation can be used as an effective teaching tool (demonstrating the need to keep the plaster moist) but only if the masons and homeowners are beyond the point of simply blaming the bales.

Plaster coats are also put on in very quick succession, leaving no time for cracks to form before the next coat is applied. Since masons are accustomed to cement-based plasters, they are unwilling to accept limitations on the thickness of layers and the time required between coats (both of which are necessary with lime plasters). For these reasons, and because of the quick drying of the plaster, cracks often telescope through all three coats.

Materials Issues

China is in the midst of a race toward modernity the scale of which the world has never seen. The hunger for newness runs so deep that it has a mildly religious quality about it. Concrete, steel, glass, vitreous tile, brick — these are the materials that make people proud. Cement, in this symbology, is the unassailable king of construction materials. Cement/sand plasters (mixed 1:3) are also what people are accustomed to using over brick houses, and it can be very difficult to convince them that a plaster which works well over bricks is not necessarily the best choice to use over bales. Earthen plasters, on the other hand, are exactly what people are trying to get away from — if you use earth that means you are poor.

Lime is generally considered to be inferior to cement, because it's cheaper. On the other hand, it is made in a factory. It is commonly used for interior work, but people are resistant to using it on exteriors, as it is thought to be less durable than cement. In a limited sense, this is accurate. The quality of lime (usually hydrated) varies tremendously from site to site, and with storage and handling conditions. All of the lime seems to be of a lower grade than that available in the West.

Conclusions

After preparing many samples of clay and lime-based plasters, we have finally realized that for straw bale construction to succeed in China, we must yield to the strong cultural bias toward cement. The acceptable compromise is a plaster in which cement is used in a ratio no greater than 1:1 with lime, and in which reinforcing fibers are used in all but the finish coat. This is a not perfect solution, but it has reduced cracking tremendously (compared with the original 1:3 cement/sand mixes). The resulting plaster can be applied as thickly as necessary in each coat, and time between coats is less crucial than with lime plasters. Most importantly plasterers need only alter their usual process a bit by moistening the walls several times a day to prevent the plaster from drying out too quickly.

PAUL LACINSKI

5-16: *Lime plaster drying on a brick and bale house. Note the differential drying rates.*

Sustainable Settlements: Rethinking Encampments for Refugees and Displaced Populations

Cameron M. Burns

The average modern North American knows virtually nothing about refugees. Yet as humankind proceeds quickly into the 21st century, this group of people could become the most deserving of our attention simply because it's likely their numbers will grow. Continuing desertification of sub-Saharan regions, climate change and rising sea levels, ongoing resource shortages, and the violence resulting from such shortages will all be felt by the poorest members of society first.

The UN estimates that worldwide it cares for an estimated 22 million refugees, but that number, agency officials are quick to point out, might represent only half of all refugees. Some refugees are dispossessed for only a few weeks or months. Others have held their status for years. Some have even been refugees for several generations.

The camps that refugees come to call "home" can be awful, which is no surprise. When disaster, war, and shortages prompt refugees to flee one place, they often do so by the thousands or tens of thousands, even by the hundreds of thousands in stunningly short periods of time. For example, during a three-year period beginning in 1990, 100,000 Bhutanese asylum-seekers fled into southeastern Nepal; between 1992 and 1997 Tanzania received 800,000 refugees from Burundi and Rwanda; and between July and October 1994, 730,000 refugees fled Rwanda for Goma, in the Democratic Republic of Congo. And in a compelling example of mass exodus, 250,000 Rwandans, fleeing ethnic violence, crossed the border into remote Northwestern Tanzania in just two days.

Under such circumstances, aid workers are pressed to erect tent cities within weeks, even days. Order must be maintained. Water, food, and clothing must all be obtained immediately; then an ongoing source for these basics must be established. Not surprisingly, the task of taking care of refugees is falling more and more on military organizations, which have the skill and discipline to deploy quickly and create order out of situations that might otherwise progress into anarchy.

How refugees are handled, and the way in which their habitations are established, is becoming of greater interest both in military circles and among aid organizations. One man who has become deeply involved with refugee camps and populations is Dr. Eric Rasmussen of the US Navy. Rasmussen's work in refugee settlements has shown him that the aid being brought to refugees can create problems as big or bigger than the issues being addressed.

"When refugee camps are set up," Rasmussen notes, "the urgent circumstances require that the basics of food, water, shelter, and safety be delivered just as quickly as possible or lives can be lost. Because the responsibilities for sectors are split across many agencies, isolated answers to a single problem are often the result. Unfortunately, despite superb efforts and many saved lives, the resulting infrastructure is often less than ideal and becomes semi-permanent. Such disconnected coordination can cause seemingly foolish problems that are invisible until you work in the camps.

"At one camp in Africa, for example, one aid agency delivered drinking water from a 2-inch (5-centimeter) pump spout while another agency provided plastic water containers with 1-inch (3-centimeter) openings. These particular refugees weren't familiar with funnels, so the simple mismatch resulted in thousands of gallons of spilt water. The spilt water created a mudhole. The mudhole was fixed when, rather than fixing the spout-jerrycan mismatch, a different aid agency laid a cement slab with a sluice leading to a shallow collecting pond for the spilt water runoff. The result was a mosquito-infested pond 30 feet (9 meters) from the water pump and a 40 percent malaria rate in those who used that site to pump their drinking water. This is a design problem."

A Refugee Primer

Organized refugee care is a fairly new phenomenon. In modern times, it was at the end of World War II — when an estimated 40 million Europeans were displaced — that the world community began looking at and understanding the plight of the dispossessed. In 1951 the UN wrote the international Refugee Convention, which defined a refugee and outlined "the minimum humanitarian standards for the treatment of refugees."

Officially, a refugee is a person who "is outside her/his country of origin (or habitual residence, in the case of stateless persons) and who, owing to a well-founded fear of persecution for reasons of race, religion, nationality, membership of a particular social group, or political opinion, is unable or unwilling to avail herself/himself of the protection to which s/he is entitled."

The problem with the 1951 Convention definition, according to David Stone of the United Nations High Commissioner on Refugees (UNHCR) and Larry Thompson of Refugees International, is that the UN definition leaves out quite a few folks, notably people uprooted within their own countries, so-called internally displaced persons (IDPs). Further confusing matters is that in places like Afghanistan — where sustainable designs might first be applied — there are "old" and "new" refugees, according to Thompson.

An estimated four million "old" refugees resulted from the Russian occupation and war of the late 1970s and 1980s; the "new" refugees have been displaced by more recent fighting and a 1999–2001 drought. In late 2001 a vast new flood of refugees was feared in the wake of US military action, but international efforts to deliver relief aid inside Afghanistan, enabling Afghans to remain in their homes, were relatively successful.

Not all refugees are created equal. The roughly one million Afghan IDPs who could not cross international borders in 2000 and 2001 (partly because neighboring countries closed their borders) don't have the same rights as international refugees, and are often aided in only a minimal fashion or not at all. Moreover, many refugees are overlooked by the main humanitarian efforts because they integrate quickly into local populations, as have many Afghan refugees who have fled to Iran and Pakistan.

The camps where refugees wind up are usually in poor nations, and they enormously burden local societies, economies, and ecosystems, leading to a swarm of problems. Armed militia and guerrilla factions sometimes infiltrate camps and terrorize refugees; violence against women, children, and other vulnerable people is common. Sometimes those hired to run the camps come from a local population that has been at war with the refugees, prompting severe mistreatment. Locals outside the camp often resent the international aid the refugees receive and steal whatever they can from the camp inhabitants.

Sometimes the refugees themselves don't trust the aid — as workers in Sudan found when refugee mothers refused to feed their starving children because they feared the food was poisoned. Refugees are sometimes inadvertently given food, supplies, and fuels that break cultural or religious mores. Sometimes they're given food that requires considerable cooking, prompting energy-related problems such as deforestation.

Even local governments can throw up obstacles. At one African camp the UN wanted to initiate several environmental projects. The national government — which had been charging rich Western humanitarian groups large sums of money simply to gain access to refugees within its borders — demanded US$20 million from the UN to begin work. The UN refused and eventually gained access to the camp, but such extortion adds one more complex problem to the mix.

According to Refugees International's Thompson, a typical refugee camp can house 10,000 people. But camps may have hundreds of thousands of residents, as was the case with Rwandan camps in the Congo in the mid-1990s — one of which grew to 600,000. Refugee camps are supposed to be temporary, but unresolved conflicts often make it difficult for refugees to go home, and the camps can remain for decades.

Rethinking the Design Issue

As Dr. Rasmussen notes, design is at the center of many refugee camp problems, but the answer might not simply be to hire new designers. Some of the issues go far beyond poor communications about projects shared by aid agencies. There are endless stories from refugee camps where well-meaning aid organizations have provided advanced technological devices, the best foodstuffs, and other new, expensive materials that simply do not match the economic, educational, cultural, and geographic realities of the situation. Dr. Rasmussen feels strongly that such situations call for an overlapping integration of players from diverse backgrounds. He thinks the sustainability community's approach of understanding an entire system before attempting a solution might be the appropriate approach in refugee settlements.

Properly combined, today's best innovative practices can often provide for basic human needs — clean water, food, sanitation, shelter, security, light, refrigeration, telecommunications, medical care, and education — in ways that support prior populations, check the spread of poverty-inducing conditions, and restore vital habitat and infrastructure. Moreover applying key insights from other disciplines can even help to create a sound sociology, an entrepreneurial micro-economy, and a sense of dignity and self-worth.

The combining of many proven solutions, normally deployed only singly, should yield very important synergies. Making the skills and techniques scaleable and portable — so that refugees can take them home to help with rebuilding — could make repatriation more likely and more successful and create a nucleus for national development. And if this can be done in refugee camps, it should also help some two billion or more other people seeking to create sustainable settlement in austere conditions.

In mid-February 2002 the Rocky Mountain Institute partnered with Dr. Rasmussen to rethink refugee and displaced persons settlements from scratch. The United Nations High Commissioner for Refugees (UNHCR); Refugees International; the UN Development Programme; the World Food Programme; the US State Department; the Departments of Energy and Defense; among many other NGOs, government departments, and individual specialists were also involved.

The event, officially called the "Sustainable Settlements" charrette (a charrette is a type of "design brainstorming" process), took place at El Capitan Canyon, a rustic camp and retreat center near Santa Barbara, California. Use of El Capitan Canyon was donated and the event generously hosted by co-owner Chuck Blitz. Other costs were borne by generous grants from private donors. Leaders from the aid community met with some of the best integrative design practitioners for sustainable development to seek ways to manage refugee settlements more effectively.

Developing Projects

So what should a nation do if, say, it was suddenly faced with a three- or four-month-long influx of 100,000 people into a community, all of whom needed immediate help? Or 200,000 people? How about half a million? The 84 attendees at the charrette formed working groups covering all the issues of concern to UNHCR — energy, site, water and sanitation, communications, education, health, economic development, food and nutrition, construction and shelter — and were assigned to envision three projects that could be implemented within six months. They were also given a location for their theoretical efforts: the community of Spin Boldak, where an encampment near the Afghan-Pakistan border formed in late 2001 with nearly 10,000 IDPs (mainly women and children) offers the possibility of using ideas from the charrette in a real-life setting. (Ideas generated from this charrette might also be applied along the US-Mexico border, in rebuilding Kabul, and in many other settings.)

Some of the results were revolutionary, Food, for example, arrives in all sorts of packaging, most of which is discarded. But boxes of aid materials could be impregnated with crop seeds and spores of fungi that help gather nutrients and hold soil. Each box panel could fit a region and season, be ready to plant, and create a kitchen or market garden just by putting it on the ground and watering it. Charrette participant Paul Stamets of Fungi Perfecti is already talking to packaging firms about making such boxes.

How about education? Such a "seed box" could deliver a "School-in-a-Box" — another charrette idea that could supply refugees with camp information, learning materials and school curriculum, gardening supplies, solar toys, solar power information, etc. Even some of the simplest but currently unapplied ideas could be helpful in camps.

"The first project our group developed was an assessment of the refugees themselves, an inventory of the human resources," noted RMI's Michael Kinsley, Economic Group facilitator. "There's a lot of brainpower that comes into these camps, and camp organizers should be tapping into that resource." Not only does an assessment provide humanitarian agencies with information about the population, Kinsley noted, it could empower the refugees themselves, by building self-esteem and getting them involved with camp projects. It also helps prepare them for their return home. And if the inventory goes on a smart card rather than an ID card, it can also represent an unstealable personal store of value (set up with microcredit upon registration) to jump-start local commerce.

An Energetic Flow of Ideas

The individual projects the charrette produced were impressive, but it was the way in which complementary knowledge and experience were connected and woven together that made this design process unique. A poignant example came from the charrette's Energy Group, which comprised technology and fuels experts, solar and adobe experts, and experienced aid workers.

On their first day group members pondered how to get the most heat and light from various fuels, and which fuels were appropriate. They came up with some good ideas, but the arrival of Afghan refugee Fauzia Assifi and an Afghan-experienced nurse-anthropologist caused the group to refine good ideas into great ones.

Afghan families, Assifi explained, are accustomed to heating their feet and lower legs by sitting together (*sandelei*) around a table, covered with a heavy quilt, with a small charcoal brazier (*manqal*) underneath — an arrangement similar to the Japanese *kotatsu*. The brazier, containing coals covered with ash, stays hot for many hours. Afghans cook, eat, and share each other's company around the manqal and often go to sleep in the same positions, leaving their legs under the brazier-warmed quilt and stretching out on their sleeping mats.

Building on Fauzia's information, the Energy Group decided that a new type of brazier insert might be in order. Fueling it — and an efficient stove/pot combination for cooking — with LPG (bottled gas) could greatly decrease the environmental damage resulting from cooking with fuel wood (and then trying to heat people with the same cooking fire). It could free up the excessive fuel wood gathering time required of women and children, so they could further their education or earn more and could avoid landmines and attackers while foraging for firewood. It could also eliminate indoor smoke and eye damage (which is chronic in Afghanistan), without many of the risks of kerosene. A trickle brazier that uses only a tiny amount of LPG could thus provide personal warmth to family groups in the evening and at night in cold climates, in a way that reinforces family cohesion and traditional practices.

The Energy Group took the discussion even further by hypothesizing that such new technology might stir the interest of gas, oil, and LPG companies (such as those now emerging in Afghanistan), which could see new markets created through technologies introduced for refugees. The discussion was rich and deep.

The roughly two dozen projects developed were then considered on an integrated basis, taking cultural and technological appropriateness and resource preservation into account. Yet as the working groups pondered their projects, it became apparent that there are several larger ideals humanitarian agencies must follow.

Emergent Themes

A number of very important themes and goals cut across all the projects and unified them in special ways. First, all charrette participants agreed that the refugees themselves should be encouraged to lead efforts to provide aid. Refugees know their cultures, religions, regions, and desires better than any Western aid worker. Having refugees lead their own efforts in all eight areas (energy, site, water and sanitation, communications, education, health, economic development, food and nutrition) not only builds esteem in the refugees, but it assures that well-intentioned help doesn't get misapplied.

Second, the help must be appropriate — culturally, religiously, economically, technologically, geographically, and in terms of resources. All charrette projects put a strong emphasis on education, both for aid workers and for refugees. Learning is the basis for all good, solid, appropriate work, and cultural imperialism is a habit we should all strive to avoid.

Third, all areas of aid should be coordinated from the start and throughout the displacement period of the refugees. Some projects meet each other across a topic area boundary — the "seed box" and "School-in-a-Box" idea is an immediate example, where the ideals of the Food Group join with the goals of the Education Group to meet a need in a sustainable way. Such coordination is extremely important (it was a lack of coordination that prompted RMI's charrette in the first place).

And finally, the projects themselves must be more fully developed. How they leverage one another; support cultural goals; and enhance the environment, economy, and lives of these dispossessed people must be completely understood before they are taken out and tried. As UNHCR's David Stone put it, "Please do not try and take any untested or unproven techniques or tools to a refugee setting, and certainly not to an IPD setting in which the infrastructure is less supporting than in many refugee camps. There's a lot of things we need to do before we can take some of these individual activities, put them together, and deliver them whole to IDP or refugee camps."

Where Do We Go from Here?

The Sustainable Settlements charrette was not undertaken to produce floor plans for camp buildings and design drawings for new cooking devices; rather, its purpose was to create a settlement design methodology and template for quickly helping displaced people — in short, a primer for aid workers.

©CAMERON M. BURNS/ROCKY MOUNTAIN INSTITUTE

Representatives from several large aid organizations have expressed interest in the outcome of the charrette, including UNHCR, UN Development Programme, the World Health Organization, the World Food Programme, Refugees International, USAID, and others. According to Dr. Rasmussen, there is also interest in using the ideas for domestic sites within the United States, in depressed or marginal communities.

The President of the Massachusetts Institute of Technology (MIT) recently mandated the university to look for "expeditionary" opportunities, find people active on field projects, and promote and support the work being done in those projects. "It might be an architect building a small community in Turkey, it might be a geologist working in Nepal," explained MIT's Mike Hawley. He noted that MIT is interested in "tugging up those projects, giving them better visibility, funding, resources, and real incentives to synthesize the kinds of skills that are needed across traditional boundaries in the university."

6-1: Sustainable Settlement Charrette participants at a retreat center near Santa Barbara. Participants included representatives from The United Nations High Commissioner for Refugees (UNHCR); Refugees International; the UN Development Programme; the World Food Programme; the US State Department; the Departments of Energy and Defense; and many other NGOs, government departments, and individual specialists.

Additionally MIT has an "entrepreneurship competition," in which projects are developed, with business plans, and entered into the competition. Local venture capitalists assemble $50,000 in prize money, examine the various projects, and fund six winning plans. Some go on to become viable businesses. Hawley felt this type of approach might be one way to get some of the charrette projects moving quickly.

The North American Development Bank is interested in experimenting with sustainable refugee camps along the US–Mexico border, where "a constant flow of refugees," according to author Alan Weisman, is present. The bank has reportedly set aside $23 billion for such activities.

The Projects

The projects developed by the eight working groups at the charrette are only briefly described here. For more information, please visit the Charrette website at: www.rmi.org/sitepages/art7205.php.

Food and Nutrition Group

The ultimate goal of the Food and Nutrition Group's work would be to get refugees to feed themselves. It is important to achieve self-sufficiency before the next emergency forces support agencies to redirect their efforts. The group focused on the following projects:

1. "Knowledge Scoop"

A "self-feeding" camp starts with good knowledge. A high quality assessment that considers resources, constraints, requirements, and relationships is vital for determining the best ecological fit for the camp. The Food and Nutrition Group therefore recommended a "holistic, comprehensive, integrated, multi-agency and full cycle response assessment" process, which they dubbed the "knowledge scoop." Both local and outside experts would perform assessment, possibly with assistance from the refugees themselves. It would create a "virtual guild" of expertise for sustainable relief by considering such things as topography, hydrology, traditional agricultural methods, capacity, human capital, coping skills, regional context, and diet. The Scoop would provide the best information on the best long-term responses to meeting refugee food needs, guarding against donor fatigue, mitigating environmental damage, and supporting self-sufficiency.

2. EcoAction Team

The EcoAction Team (EAT) would coordinate information and implement recommendations coming from the Scoop — offering resources and expertise for camp inhabitants. It would be drawn from and serve as a resource to multiple relief agencies as well as to camp residents. EAT would be both a group of people and a physical center. Its purpose: to increase camp food production by linking the emergency (Phase One) food delivery system to an evolving food production (economic) system, promoting local food production expertise, helping to turn all camp inputs into resources, and teaching the teachers. The physical delivery point for emergency food would become a learning and community development point. Along the way the EAT would help to monitor the overall health of the camp and work to improve it, create a food knowledge base and process, promote natural capital, and support camp governance. EAT should appeal to food aid providers, since increased self-sufficiency would reduce demand for outside aid.

3. A Box to Save the World

The highlight of the Food Group's projects was the "Box to Save the World." In the Group's vision, all debris flows at the camp — and in particular food packaging —

would be turned into soil, seedbeds, and other supports for food production, habitat improvement, and self-sufficiency. The "Box to Save the World" would be the same box typically used during the emergency phase to distribute food rations. The difference would be that this box could itself grow into more food. To start a garden, recipients would simply spread the box on suitable ground and add water. Manufactured of highly biodegradable material, the box would be impregnated with seeds of appropriate foodstuffs (or other useful plants), plus mycorrhizal fungi to help the seeds take root. Any box destined to reach the camp could be impregnated with seeds and agricultural products to provide a livelihood for refugees and help reverse environmental degradation. Seeds would be selected from naturally occurring, region-appropriate and season-appropriate plants, including annuals and perennials. Use of the boxes would develop transferable human and physical capital. Creating this "implement" would support cottage industry and would also be an excellent candidate for private sector partnering.

Water and Environmental Sanitation Group

Water is generally the most pressing immediate requirement for refugees, and the most important ongoing one, as well. Part of the water challenge is that most refugee settlement locations don't have a significant water supply, so it must be trucked in. And what is trucked in isn't that clean — just sucked from a nearby town or pond.

The establishment of a safe water supply, therefore, was the first objective the Water and Sanitation Group identified. Their first two projects represent the importance of a clean water supply. Their third represents a recycling of water.

1. Mobile Emergency Relief Water Treatment System

The technology already exists to manufacture portable treatment systems that can provide an immediate, safe water supply from almost any source. Using their knowledge of available technology, the Water and Sanitation Group partially designed a water treatment system.

2. Improved Sand Filters

A longer-term supply of safe water is critical to camp stability. The group thought that its second most important project should be a longer-term treatment system and that upgrading existing UNHCR sand filters with low-power, ultraviolet disinfection technology could achieve that goal. Furthermore this non-chemical method would be superior as a sustainable solution.

3. Reed/Wetland Wastewater Treatment

Since wastewater is unavoidable, it might as well be used for something. The group envisioned using wastewater for agricultural irrigation. Wastewater would first be run through a reed bed/wetland system for treatment, before being applied to crops and trees. The capital costs of creating such systems would be extremely low, and the results are already proven (currently there are more than 5,000 systems in operation worldwide).

Energy and Energy Supply Group

The Energy and Energy Supply Group was charged with figuring out appropriate systems for cooking, warmth, and light and for identifying appropriate fuels. The group developed three projects.

1. Fuel and Technology Package for Cooking

The group thought that alternative fuels should be explored, since the use of fuelwood has negative societal and environmental consequences; liquified petroleum gas (LPG) was one possible candidate. (Kerosene, commonly used in camps, is very dangerous and causes carbon monoxide poisoning.) Because cooking requires 70 percent of camp energy, a reexamination of LPG cooking devices was suggested. The group also suggested that the cooking devices ordinarily distributed to refugees (pots, pressure cookers, kettles, etc.) might be more efficient if they were made of better heat-conducting metals.

2. Communal Warmth: Propane Trickle Brazier

In many camps in cold climates (such as Afghanistan) there is a tremendous need for personal warmth for family groups in the evening and at night. The energy group wanted to find a way to deliver these things, while at the same time reinforcing family cohesion and traditional practices. So they proposed the development of a trickle brazier with the capacity to use two fuels — charcoal or LPG with a catalyst burner. Such a device would cost considerably less than two separate devices and would have many side benefits, as well (no kerosene smoke, better efficiency, etc.).

3. Personal and Security Lighting

Lighting is necessary for both personal and security reasons. The group thought that individual lighting — run on solar or other-method rechargeable batteries — could easily be delivered by LED lights. Larger area lighting could be powered by solar rechargeable batteries/off-grid, with overlapping coverage to eliminate gaps/dark spots. Schools and community buildings could be daylighted.

Communications Systems Group

The Communications Group did not approach communications as a goal in itself. It was discussed as a supporting component of other activities, such as commerce, education, political involvement, and other activities. Yet communications still need to be developed in a sustainable manner — low power, off-grid, recyclable, economical, and easy to use.

The group developed three projects.

1. "Re-Purposed" Toys

Adaptations of technology can offer tremendous benefits to displaced people. Since it's always children who adopt gadgets first (followed sometime later by adults), a device

©CAMERON M. BURNS/ROCKY MOUNTAIN INSTITUTE

that would appeal to children was the first project discussed by the Communications Group. Specifically, they envisioned a small, personal, low-cost device ("toy") that could receive a signal or read programmable material. (Personal transistor radios and the more modern Walkman™-type devices are examples of such devices.) Extremely cheap in the West, these devices use cards, chips, or tapes to carry information, or they can receive radio signals. Re-deployed for use in camps, they could offer everything from information about the camp to culturally appropriate programming. After the distribution of personal devices, aid workers would need to follow up in two areas.

6-2: A small satellite dish demonstrated at the charrette is an example of an inexpensive, high-impact technology that could be used to improve the lives of refugees.

2. An Information System ("Camp Radio")

Voice (and eventual text/Web) communications could provide tremendous support for all camp activities. Camps would need a broadcasting or information source to supply information and entertainment to camp inhabitants — in essence, the broadcaster who sends a signal or disk to the "Re-Purposed Toys." A local GSM-based information system, along with a regional satellite-based system could provide the content needed to inform, instruct, and entertain refugees.

3. Telecomm Education Centers

At a telecomm education center, refugees and IDPs could learn the technical aspects of communications technology (how to use the personal devices and how to run the information services), as well as the myriad ways that communications can bolster commerce, education, and other activities. The training center should reflect whatever tools and infrastructure already exist in rural areas, so that people being trained could continue to expand communications services once they return home. Training people to develop economic models that promote commerce in rural areas is an important component of this three-part project.

Health Group

The most important factors that influence health in a refugee camp (as in society at large) are outside the usual jurisdiction of doctors. Thus an integrated health policy is critical to maintaining healthy standards in refugee camps. Doctors need to be involved in the larger issues of refugee health, including monitoring community systems and community health (regularly checking water purity, monitoring child malnutrition, etc.) to catch causes of health problems before they result in epidemics. Typically by the time a doctor gets involved with refugees, the state of their health has already been determined by the sanitation, nutrition, security, economic stability, and mental health support network in the camp.

1. Family Planning

Family planning is critical to both human and environmental health. It is such a culturally sensitive issue, however, that it is absolutely essential that natives of the culture provide the service, and provide it in a culturally appropriate way. What might work very well in some cultures (cartoon characters that promote contraception in the Far East, for example) won't work in other cultural settings (such as Afghanistan). Aid agencies should work with the local community and religious leaders to assist them in providing this service. Women's education is likely the key to family planning: education will not only reduce child mortality rates but will also increase family health at low cost. (A study on longevity found that political commitment, female literacy, nutrition, and equity healthcare were common factors in the five countries that most successfully achieved longevity at low cost.)

2. Mental Health Treatment

Post-traumatic Stress Syndrome, depression, apathy, and boredom are the main mental health problems in refugee camps. (Mental health disorders are as debilitating as tuberculosis.) Good mental health is arguably the best asset of the camp; restoring it improves the capacity of the whole community. Building a community facility that provides meeting space for peer support would be of tremendous benefit to people with Post-traumatic Stress Syndrome.

3. Nutritional Supplements

Since the most effective way to promote health is to prevent disease in the first place, the group recommended that camp inhabitants receive key nutritional supplements to boost their immune systems. As did the Food and Nutrition Group, the Health Group turned to mushrooms to most effectively and inexpensively serve this purpose. Seven basic medicinal mushroom varieties could be incorporated into the "Box to Save the World."

Education Group

Although UNHCR defines education as one of the basic rights of a refugee, few camps provide formal schools. If one does exist, it is often an informal gathering under a tree or in a corner of the camp. The Education Group proposed that education should not be the caboose, but rather the engine driving refugee settlements for all members. Key components would include: a local vision to create and move initiatives forward; sensitivity and inclusiveness, particularly regarding women's issues and illiteracy; assessment, monitoring, and evaluation; and ongoing support for programs. Using these requirements, the education group identified three modules.

©CAMERON M. BURNS/ROCKY MOUNTAIN INSTITUTE

1. Train the Trainers in Sustainability

The "train the trainers initiative" is intended to empower the "community animators" who are present in every group and insure that education of refugee populations includes vital community-building skills and sustainability training. This module would provide support and training to help educators pass on their knowledge to the community as effectively as possible.

6-3: Charrette participants brainstorming solutions to the problems faced in refugee settlements.

2. "School in a Box"

This initiative is intended to provide basic materials, how-to information, physical capacities, and curriculum content in a large physical box, for the purpose of establishing both a basic school and additional programs. The "School in a Box" concept is already in use in some places (UNICEF, Rishi Valley). The box should contain both conventional tools and learning materials (books, paper, pencils, etc.) and interactive materials designed to encourage experiential rather than rote learning.

3. Community, Life, and Repatriation (CLR) Skills

CLR skills are intended to empower refugees with the skills to become independent, self sufficient, and prosperous upon return to their homeland. Refugees must be able to lead and rebuild their communities both within the camp and upon their return home. The initiative would be designed to cultivate a set of practical skills that combine indigenous resources and expertise with best-of-breed techniques for building positive and sustainable community elements: gardens and farms, homes and businesses, clothing and paper, etc. In addition to vocational skills, this module seeks to help rebuild an indigenous system of justice by supporting community leadership and self-regulation within refugee settlements.

A school within a refugee camp should be a center for all modes of learning, where people of all ages, both male and female, can get apprentice-style, hands-on experience to learn the skills to build a practical livelihood and a healthy community (weaving, masonry, and adobe-making, for example). Resources for the development of this module included the Gaviotas model, the Peace Corps, and Sustainable Village.

The Economic Development Group

The Economic Development Group confronted a daunting dilemma: how do you improve the refugees' economic situation without encouraging them to remain in their camps? The group concluded that economic development efforts must be focused on building the self-sufficiency of the camp economy to help make life in the camp more bearable and reduce the demand for relief services; and on strengthening refugees' economic skills and potential, so that they are better prepared to rebuild their home economy when they return. The group quickly learned that, in general, relief workers have insufficient resources or capability to provide economic development to refugees. Thus the group's first suggestion was that an inventory be made of a camp's human resources.

1. Refugee Skill Inventory

Despite outward appearances and wretched conditions, many refugees have valuable skills that can be put to use in camps, both directly and in teaching fellow refugees.

Though camp registration in the early emergency phase of camp life is difficult, even chaotic and dangerous, the Economic Development Group was convinced that camp organizers should devise culturally sensitive ways to inventory refugee skills and add skills-related questions to registration protocols. Putting those skills to work in the camp in an organized way would increase the dignity, self-respect, and economic potential of residents while in the camp and upon repatriation.

2. Development/Business Center

Group members envisioned a center that would teach and support skills (farming, crafts, business, management, leadership, etc.) to improve conditions in the camp (and even in nearby communities) and strengthen the capacity of refugees to rebuild their home economies upon repatriation. Such a center could also provide micro-credit and technical assistance for fledgling businesses. Refugees, to the extent possible, plus a new cadre of international development workers could provide classes and other services. The center could be integrated with centers recommended by the food and communications groups.

International Development Workers Training Institute

Although relief workers are extraordinarily committed and energetic, few are prepared to help refugees develop their economic potential. Therefore the Group proposed the formation of an institution to prepare relief and development workers to support refugees in developing stronger camp and home economies. This institute would develop operational understanding of both social and ecological restoration activities. Primarily Web-based, it would also include on-site experience in delivery of camp development/business centers.

Site Planning Group

Although much information about helping refugees exists, it exists in pieces in different places. And it leaves out a great deal of socio-cultural information about the refugee population itself. The Site Planning group suggested three projects.

1. Information "Reachback" Project (Database)

Humanitarian groups often find themselves trying to establish campsites without any background information. To alleviate that problem an information database could be assembled and made available to humanitarian workers on the ground and to workers in training, with information directed at three challenges: training, problem solving, and strategic decision making. Information could be shared via the Web, paper hardcopy, portable electronic forms, etc., with the database itself housed with UNHCR. Such a database would benefit all phases of camp management. Information could be shared with universities and other learning institutes, as well as with funding groups.

2. Socio-cultural Information Project

To ensure socio-cultural sustainability in refugee camps (alongside environmental and economic sustainability), the Site Planning group thought that a "socio-cultural information gathering project" was important. Similar to the refugee assessment described by the Economic Development Group, this project could focus on cultural aspects of the displaced population rather than on individual skills. Field researchers could interview and observe refugees to gather information about how to best develop, change, and operate relief efforts — everything from how a building should be sited to which activities could appropriately occur next to one another. Aid agencies could gather and distribute information on socio-cultural factors specific to a region or group, sharing the results with both refugees and aid workers. "The point to remember here is that the refugees themselves are the experts," noted Clare Cooper Marcus (see "Site Planning For Refugees: What to Ask").

Strategic Operations Planning (For Site Selection)

At present humanitarian relief efforts occur in reaction to events; we do not plan and prepare for them. Global hot spots, where historical conflicts and geo-politics are likely to lead to the mass displacement of civilian populations, should be acknowledged and planned for. (The group thought that at any given time, there are 20 to 30 such places around the globe.) Such a strategic plan should take multiple forms and be available immediately when crises occur. The plan would need to be updated as the project is progressing.

Charrette participants will continue their healthy, rich dialog and share it with whatever other individuals, organizations, and governments are interested. Unfortunately the future of refugee camp business is strong. As the World Health Organization has noted, "almost two billion people — one-third of humanity — were affected by natural disasters in the last decade of the 20th century. Floods and droughts accounted for 86 percent of them." Add to that coming climate change, future wars, and resource shortages, and it becomes apparent that the demand for clean, healthy, habitable, and sustainable settlements is going to go up, and not down.

SITE PLANNING FOR REFUGEES: WHAT TO ASK

Clare Cooper Marcus

Refugees and displaced persons can provide a wealth of information to the site planners of temporary or semi-permanent camps. But planners need to know what to ask. The following questions are designed to help that process and are deliberately generic, making them useful for differing socio-cultural situations. Assume that each question is prefaced by the words, "In your culture/tribe/clan/group...." (Also see "Sustainable Building As Appropriate Technology" and "Down to Earth Technology Transfer.")

1. What does a typical family consist of (mother, father, children, uncles, aunts, grandparents, etc.)? What size might this family be?
2. Do all the members of a family sleep in the same space?
3. Are sleeping arrangements different in hot/cold or wet/dry seasons?
4. Do you traditionally group or cluster dwellings? If so, how many live in a cluster?
5. How are cluster dwellings oriented? Are they back to back? Side by side? Do doors or windows connect? When dwellings face each other, are entries offset to ensure privacy?
6. How does the cluster relate to public space, such as a footpath, street, or alley? (Is there a narrow, controlled entry into the cluster? Are all dwellings in the cluster oriented to open to public space?)
7. What materials are used in the construction of a typical family dwelling?
8. Are there any materials that would never be used to construct a dwelling? Why?
9. When constructing a dwelling does the entry or any particular room traditionally face a certain direction?
10. Is there a particular space that is used predominantly/exclusively by men/women?
11. What is typically stored inside the house?
12. Traditionally are there some spaces in the house that are private (only used by family) and some that are open to guests?
13. In the construction or decoration of a dwelling, are there any rituals or practices that transform a dwelling into a home?

14. Is there a place just outside the dwelling where family members might sit and socialize? How would it relate to the interior (for example would it be just outside the front door) and to nearby public space (would it be close to passersby on the street)?

15. Where does a typical family eat (around the fire, inside, on the floor, at a table)?

16. Do men and women eat together?

17. How and where is cooking done? Does it differ in winter and summer? In wet and dry seasons?

18. When cooking do families in a group/cluster all use one fire? If there are separate fires, how are they oriented in relation to each other?

19. For how long each day is a fire typically kept alight for cooking?

20. Are any foods (such as bread) prepared communally?

21. What fuel is typically used for cooking, and who collects it?

22. Is the same fuel used for warmth?

23. When warmth is necessary how is it provided? By heating a room? By putting on more clothing? Or by warming a part of the body (feet, for example)?

24. What are the essential crops used for food?

25. Who is traditionally responsible for food production (men/women/boys/girls)?

26. Are there some foods that are traditionally grown right outside the door?

27. Are there any traditions of drying or preserving food?

28. What foods or medicinal plants are traditionally gathered in the wild?

29. In a traditional settlement, who usually collects water, and how many hours per day are spent in this task?

30. What are the potential hazards encountered by those collecting water or fuel (snakes, wild animals, rape, exhaustion)?

31. Where are the latrines traditionally located?

32. Do men and women use the same latrines?

33. Do you use human excrement to fertilize the ground?

34. How and where do men/women/children bathe?

35. How and where are clothes washed?

36. Traditionally are there outdoor space or buildings in the community predominantly/exclusively used by men/women?

37. Is there a traditional place for prayer or worship? Are there traditions regarding its location?

38. What are a man's/woman's responsibilities in and close to the dwelling (cooking, cleaning, child-care, food preparation)?

39. What are a man's/woman's responsibilities away from the immediate family dwelling (livestock, hunting, gathering, collecting water, finding fuel, raising crops)?

40. At what age do a man's/woman's responsibilities begin?

41. What are the income-producing skills of men/women/children? What raw materials/work environments/marketing arrangements are necessary?

42. What space/environments/equipment are necessary for the desired recreational needs of men/women/children?

43. Are there any measures/traditions for dealing with trauma or mental illness?

44. How is sickness and ill health handled?

45. How is a man's status measured (Number of children? Number of male children? Amount of land owned? Number of cattle or pigs, etc.)?

46. How is a woman's status measured?

47. How is justice administered?

48. What will help you feel most "at home" in a camp?

Questions for Representatives of Host Community

1. What is the optimum location for a school/clinic/administration building/well in the refugee camp (to ensure its usefulness after the refugee population has been repatriated)?

2. What is the optimum location for a public space (associated with refugee registration, food distribution, etc.) that might also function as a marketing/trading place for camp residents and nearby communities?

Case Study:
Grupo Sofonias: Knowledge in the Hands of the People
Kathryn Rhyner-Pozak

Two couples, one from Latin America and one from Europe, formed the nonprofit Grupos Sofonias after the Guatemalan earthquake of 1976. We quickly learned that we were really dealing with the permanent social disaster of poverty and not just the physical disaster of the earthquake. We knew we had to concentrate our efforts in order to make a lasting difference and decided to focus our attention on low-cost, sustainable housing. (Inadequate and insufficient housing is part of the permanent social disaster throughout the "developing" world.)

For more than a year we lived off our savings until projects were approved. Then things began to move, grow, and evolve. Our early work took us to Nicaragua (after the revolutionary war ended in 1979) and to the Dominican Republic (after Hurricane David, also in 1979).

Our motto has always been "Small is beautiful." Small and simple beginnings can achieve widespread results, whether its schoolhouses built by a group of farmers or a few pilot plants built to produce tiles. After the first steps, others begin to follow suit and seek involvement in the work. It is a process that involves teaching, the creation of living examples, and projects that encourage people to take their lives into their own hands and change their living environment.

The multitudes can only have reasonable houses if they use locally produced construction materials and do the building themselves. Therefore our work is based on community participation and the use of locally produced, ecological, and economically viable construction materials that can withstand regional natural disasters. We call them "ecomaterials," and the goal is to construct sustainable dwellings or "EcoHabitats."

6-4: *Participatory planning, a hallmark of the projects of Grupos Sofonias, leads to cultural appropriateness, effective project handover, and community empowerment.*

Local Materials

Locally produced construction materials can help ease the housing deficit and stimulate local economies. Not only is money kept circulating in the region through the creation of jobs and sales of a needed product, but the transfer of knowledge and skills necessary to produce the materials also contributes to sustainability.

Currently the housing deficit in southern countries is increasing steadily, as is migration to urban areas. Yet imports from urban centers or from abroad remain the order of the day. Such a system, where housing is a commodity, can never satisfy the tremendous needs of billions of people. But if we decentralize production and place knowledge in the hands of people, they become the designers of their own destinies. Local families, masons, and builders must all actively participate, though, if the culture of construction is to change.

And it can be done. In one rural construction project in Nicaragua, for example, some 400 classrooms were built in approximately 100 villages over a ten-year period. Community participation was tremendous and involved literally thousands of people — parents, neighbors, teachers, and students.

Community Participation

Community participation is best achieved where some type of organizational base already exists. The secret is to begin small and allow the project to resonate out into neighboring villages. (In our work in the Dominican Republic, for example, requests for assistance eventually came in from agricultural associations in more than 20 villages.)

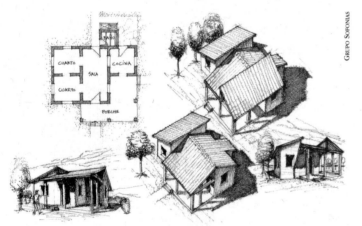

6-5: *Drawings of a well-designed, modest adobe house with a micro concrete tile roof.*

Our success has depended on the organizational mechanisms of the people involved. A steering committee, composed of representatives of various ongoing projects, would meet biweekly to organize activities for the approaching weeks. No one could hide unfulfilled tasks from their peers. Thus the project directors were not preoccupied with the many small, daily organizational difficulties that naturally arise and could concentrate on the bigger picture.

The participatory approach proved to be a great management and educational strategy. Together we built some 300 houses and re-roofed approximately 600 houses in the Dominican Republic. In Nicaragua participatory housing participants built more than 300 houses, the most recent of which was a resettlement project undertaken in Malacatoya after Hurricane Mitch devastated the area. And in Andean villages in Ecuador several hundred families have benefited from housing and roofing projects.

Micro Concrete Roofing Tiles

Early in our work we became aware that deforestation and imported construction materials were two basic problems affecting construction in the countries in which we were working. So we began to seek solutions for roofs that would require less wood and substitute for imports (see tech box on page 160).

We began to experiment with handmade fiber-concrete roofing sheets. Soon after the first roofs had been laid, however, we discovered that not only was it difficult to obtain a perfect fit between the sheets, but they were also fragile and difficult to handle. Cracks appeared over time, and we realized that the sheets were not feasible for our purposes. Clearly the size had to be reduced and some sort of mechanical process developed to guarantee consistency of size and fit.

We attended a seminar on hand-made roofing sheets in Switzerland and, discouraged with our results, prepared to tell them to forget everything. Then another participant made a presentation, showing how to use a small vibrating machine and polyethylene mold to make roofing tiles. We were in business! We bought a machine and began production.

Convinced that micro concrete roofing (MCR) tiles were an ecologically and economically viable solution for roofs, we looked for project financing. And from modest beginnings of 15 pilot plants in six countries, MCR has spread throughout Latin America. Now about 650 workshops produce some 32 million square feet (3 million square meters) of tiles annually. That translates to about 50,000 roofs each year. These small businesses do their own financing, and the project has been able to mobilize additional outside investment.

Over the past decade it has been possible to develop a network of producers, experts, and university research institutes, and activities have expanded to include other ecomaterials. The EcoSouth Network is the vehicle for the dissemination, education, quality control and investigation of MCR. Its experts provide advice and introductory and follow-up courses well beyond Latin America, into Africa, Asia, and the ex-Soviet Union.

The EcoSouth Network

The EcoSouth Network is an idea that had simple beginnings. Back in 1991 at the first Latin American MCR Seminar, participants from 11 countries decided to stay in touch, intensify the dissemination of MCR technology, and seek means to ensure a good level of quality. Since then the EcoSouth Network for an Ecological and Economical Habitat (managed by Grupo Sofonias) has guided the spread of MCR in Latin America. Since many of the key players also worked with other technologies, the ecomaterials concept quickly gained a higher profile, and the network embraced the idea of the EcoHabitat.

EcoSouth publishes a bulletin in Spanish and English, and maintains contact with organizations, producers, and other interested persons in many countries. Its experts have provided the information transfer for setting up MCR workshops and projects in countries as far afield as Tajikistan, Namibia, Zimbabwe, and Vietnam. In 1998 it held the first international Ecomaterials Conference, which attracted participants from Latin America, Asia, Africa, Europe, and North America. Its networking at the university level includes working groups concerned with pozzolanic cement, adobe, plasters, lime, etc. It also promotes bicycles as a basic means of transportation, and the planting of trees to ensure an ecologically sustainable habitat.

Disaster Prevention

The permanent social disaster of poverty is occasionally aggravated by the natural disasters — some annual, some periodic — that plague most of the Southern Hemisphere. Unfortunately most relief and reconstruction actions involve imported materials used in a conventional manner. For over three decades, we have built structures that can withstand local climate conditions. But it has been a long and hard road to convince others to integrate disaster prevention into their building projects as a matter of course. It is our dream to influence development projects in this direction.

Education is the key. Correct construction techniques, designs, and materials contribute to disaster alleviation. But participation of the people is equally important. Successful projects must integrate three community activities.

- People need to learn to build with proper designs and materials
- They need to produce these materials themselves
- The need to become aware citizens who assume responsibility for their own lives.

Through a participatory design process and on-site building instruction, people learn about risks and how best to use construction materials.

The combination was used successfully in a project in Baja Verapaz after the Guatemala earthquake of 1976. The earthquake killed some 24,000 persons and destroyed more than 250,000 houses, and the national government was considering a ban on adobe, the traditional clay construction technology, believing it to be unsafe. But we successfully used adobe to reconstruct some 150 houses in villages strewn throughout Baja Verapaz. We combined popular knowledge, the basics of engineering, and the results of a UN study to create sustainable, earthquake-resistant, and affordable housing. (The UN study, conducted after the 1970 Peruvian earthquake, identified errors in adobe construction techniques and offered workable solutions.)

Twenty-four years later a study found that the houses were in generally good condition and had withstood several tremors of considerable force, including recent earthquakes that had devastated neighboring provinces in El Salvador and Mexico. The study also found that architecturally the project was a success: it is well accepted by the people, and the architecture integrates well into the local culture and environment.

In another example, Hurricane Mitch (1999) wreaked havoc throughout Honduras and Nicaragua, necessitating tremendous reconstruction efforts. While many programs became bogged down in bureaucracy, the construction of a new community in Malacatoya, Nicaragua proceeded at a lively pace. This was possible because of the established presence in the zone of two reputable NGOs (Casa de los Tres Mundos and Grupo Sofonias) who decided to work together. And available financing made it possible to quickly purchase land for resettlement. Once construction began, another essential element became evident — the high involvement of women in building their houses for a new community.

6-6: *Community members building a house, using adobe bricks.*

The union of forces confronted the situation of reconstruction in a creative manner. Grupo Sofonias had been active building schools, houses, and community and health centers in the region for a couple of decades. The Casa de los Tres Mundos (House of the Three Worlds) had been a local cultural center for some ten years and was able to weave the local relationships necessary. The work became a lively process that integrated various local organizations (from the Rotary Club to municipal authorities and local banks) and created the conditions that made it possible for people to build their own houses. EcoSouth experts designed the new community, drawing on ecomaterials for actual construction. Families (selected according to their need and probable danger from flooding) built their houses together in small groups, with direction from an architect or engineer. Unlike most such projects, with their monotonous repetitive designs, the resulting community of Malacatoya is an attractive settlement.

Knowledge in the Hands of People

Without the involvement of people nothing is sustainable. Whether it's women from different villages coming together for a women's encounter, an agricultural association organizing construction teams for an upcoming week's work, or a whole community getting together to gather stones, organization, responsibility, production, and education all contribute to a sustainable habitat. The secret lies in placing knowledge in the hands of the people.

MICRO CONCRETE ROOFING TILES: SMALL-SCALE ENTERPRISES CREATE LOCAL JOBS

Kathryn Rhyner-Pozak

Micro Concrete Roofing tiles (MCR) are an established roofing alternative in local market regions throughout Latin America, as well as in some parts of Africa and Asia. The use of MCR tiles helps reduce roofing costs, create jobs, and alleviate chronic housing deficits, while catering to the needs of a local clientele with an elegant, enduring, and economic roofing material.

6-7: *An example of a micro concrete tile roof. These roofs are watertight, durable, relatively inexpensive, and desirable in local communities.*

MCR History

- MCR tiles replaced earlier experimental fiber-concrete roofing sheets.

- Tiles were made using specific ingredients, precisely prepared; molds into which the ingredients were poured; and a vibrating action to condense the materials and eliminate air pockets.

- The tiles solved many of the technical problems encountered when working with roofing sheets. The mixing process, vibration, and excellent curing resulted in strong lightweight tiles that were easier to make and install.

- Grupo Sofonias (see "Grupo Sofonias") quickly adopted MCR technology, and several equipment producers have since developed acceptable equipment to serve the growing number of MCR workshops.

- Various dissemination projects have introduced the technology throughout Latin America, Africa, and Asia. The impact of MCR varies according to the density of workshops in any given region and the strength with which the technology enters the mainstream housing market.

- Job creation has been a key element in most programs, and worldwide monitoring reveals that close to 8,000 jobs have been created to date, about half of which are in Latin America.

Local Impacts

- Where clusters of workshops emerge, the technology obtains a visible share of the roofing materials market.
- More jobs and the local purchase of raw materials circulate money, and local economies become more dynamic.
- Ecological and economic benefits occur because transport distances are reduced (especially important in areas requiring travel for long distances on difficult roads). Client surveys in Honduras, Ecuador, and Guatemala reveal that most manufacturers sell their tiles within a 12-mile (20-kilometer) radius of their workshops.
- Most workshops have one or two operating units; many are family businesses. In Latin America alone there are well over 600 workshops. In Guatemala some MCR workshops have flourished for more than a decade.
- Total accumulated world production is now around 308 million square feet (28 million square meters) — the equivalent of some 450,000 roofs.
- In Latin America alone, the construction of approximately 350 thousand MCR roofs has created over 3,000 jobs (a testament to the great job creation potential in the construction materials market.
- The current yearly production capacity in 14 Latin American countries is approximately 32 million square feet (3 million square meters) — equivalent to about 50,000 roofs each year.

MCR Tile Plants: An Opportunity for Enterprise

- Labor-intensive MCR technology uses locally available raw materials (sand, cement, and water). Once an initial investment is made, raw materials and labor can be readily obtained, creating a business venture particularly suited to small, informal family enterprises and people with a good instinct for commerce.
- Jobs are created for unskilled workers.
- In general women make excellent tile makers, making MCR technology a viable entrepreneurial option for this often-underemployed segment of society.
- Most tile businesses in Latin America are in the informal sector. These enterprises create jobs not only in tile production itself but also in related areas from roof laying to delivery services.

GRUPO SOFONIAS

6-8: A tilemaker in Ecuador. Making tiles has proven to be an effective strategy for micro-enterprise: building materials are produced locally and money is kept in the community.

GRUPO SOFONIAS

6-9: Grupo Sofonias offers courses in tile making and roof construction throughout Central and South America, as well as in Africa and other regions.

Producing the Tiles

Making MCR tiles looks easy — and it is! With the right equipment and know-how, just about anybody can produce good tiles if they stick to the rules and work carefully. MCR is a well-developed technology, however, with production guidelines, norms, and standards; taking shortcuts usually results in low productivity and mediocre quality.

MCR tiles have been tested against most leading standards for concrete tiles: ASTM (US), British Standards, South African, and Russian (formerly Soviet) standards. They have even been tested for freezing resistance according to DIN (Germany). The tiles have passed all generic requisites such as impermeability, bending strength, etc. And EcoSouth has its own set of standards.

- Micro concrete is defined as a "high performance concrete," which is why a ½ - inch-thick (1-centimeter-thick) MCR tile will pass tests similar to those for a common industrial concrete tile that is much thicker and heavier. To achieve this quality, an exact mixture of cement with well-graded sand, adequate vibration, and near-perfect curing conditions are essential.

- Only good equipment will produce perfect tiles. The vibrating table has to be level to permit exact preparation of the tiles, but precise molds are the decisive factor in producing correctly shaped tiles that will result in a tight fit on the roof.

- Proper conditions are achieved relatively easily and inexpensively with minimal infrastructure. A roof, a small storage area for cement and tools, a water tank to cure the tiles, and a small yard to store sand and the finished tiles is the minimal requirement for a manufacturing facility. Some people producing in their backyards have invested less than US$500 for infrastructure.

- The major weakness with the technology is how to guarantee quality when so many production units exist in so many different environments. In Latin America, the EcoSouth Network is working on establishing an EcoSouth Certificate of Quality.

Case Study: "Vaccinating Homes:"
Community-Based Disaster Mitigation in Central Vietnam

John Norton and Guillaume Chantry

In the past ten years or so poverty in Vietnam has decreased dramatically as the country emerges from decades of strife.[1] A significant proportion of families, however, still live below the poverty line and yet more live precariously just above it. Many of the latter families are "temporarily poor," resorting to a variety of activities in order to survive with irregular incomes.[2]

A house is one of the largest investments most families will ever make, requiring years of enormous effort and saving. But, cruelly, families all too often see this incremental investment destroyed by typhoons and floods. Such destruction occurs largely because "modern" materials are not well used, and simple ways of making buildings more resistant are not applied.

The risk of destruction is ever-present. Major disasters occur in Vietnam at least every decade. And the coast is hit annually by tropical storms at a rate of 4 or 5 a year or more, causing extensive and often repeated damage to housing and infrastructure, as well as losses to agriculture and fisheries.[3]

Although it is more difficult to provide protection against massive typhoons such as those that hit central Vietnam in 1985 and again in 1997, damage in the more frequent annual cyclones can and should be largely avoided. Preventive action to increase the resistance of buildings to typhoons, high winds, and floods would cost a fraction of the economic and social costs of reconstruction.

Development Workshop's (DW's) experience of working with hundreds of families in Thua Thien Hué Province since 1999 has shown that for an average house construction cost of US$980, as little as 15 percent more would make the building typhoon resistant, and this figure rarely rises to 30 percent.[4] The same applies to communal facilities such as schools and markets.

Today this additional investment is even more crucial than in the past. Paradoxically the very real improvements that have been made in building have contributed to increased vulnerability to loss — because more time and money has been invested in the home, the cost of losing it has also increased dramatically.

Fifteen years ago things were different. Village housing was typically a thatched roof, a pole or bamboo frame, and bamboo mat walls. Many of these materials could be gathered locally. Although many houses were frail and easily destroyed by typhoons, recovery could be achieved at relatively low cost. Once the immediate effects of a typhoon had subsided, village reconstruction took place quickly.

By the mid 1980s this was changing. Families were purchasing coveted "modern" materials such as cement, fired bricks, and roof sheets. But they neglected to maintain many of the storm resistant features of traditional housing. Why? The answer appears to be fourfold:

- Ignorance: families simply did not know how to make their new house stronger
- Poverty: people were forced to cut corners
- The mistaken belief that "modern" materials are inherently stronger
- An incremental building process that left many homes unfinished for long periods of time, leaving them are weak and exposed.

Today a typical family with a monthly income of US$20 to US$30 will live in a house with a floor area of 430 square feet (40 square meters) and a terrace, representing an investment of between US$650 and US$1,000. The family may well have made their own cement tiles and blocks. But they will have bought most of the other materials including cement and steel, clay roof tiles or roof sheeting (if these are used), and doors and windows. They will often have borrowed money to cover the building costs, usually at extortionate rates from informal sources. Their vulnerability has in effect increased. (Vulnerability means that it only takes a small mishap for a family to return to conditions of near starvation, ill health, and debt.)[5]

The scale of this problem is huge. In the coastal area of central Vietnam, DW estimates that some 70 percent of housing has been replaced or renewed over the past 15 years. But the same proportion — about 70 percent — of housing stock can only be classified as "semi-solid" or "weak," and thus is very vulnerable to damage. (Repeated damage creates a vicious cycle as scarce resources have to go even further.)

What Can Be Done?

In the face of repeated natural disasters, the government and the international community has had to respond both to emergency situations and longer-term rehabilitation needs. But given the scale and the repetitive nature of the damage, support for reconstruction is inevitably limited and cannot reach out to a sufficiently large proportion of the population. Meanwhile, faced repeatedly with major losses, families and local communities have to take on most of the burden of rebuilding, with only occasional external support.

In this context an ongoing DW project encourages and enables people to take preventive action to safeguard their homes and small local public facilities. The aim is to encourage people to incorporate typhoon resistant details into both existing buildings and new construction. The project's slogan is, "Vaccinate your home."

How the Project Works

We encourage families to apply ten relatively simple and inexpensive "key points" of storm resistant construction. We use animation and awareness-raising to get the message across; build sustainability by putting a local structure in place to gradually take over from the project; and provide practical help through demonstration, training, and financing.

Getting the Message Across

Most families do not know that preventive strengthening is possible. They need to learn that prevention is essentially easy and affordable and much cheaper than rebuilding after a disaster.

DW works on the principle that a serious message will be more easily transmitted and remembered if it is also fun! So the project organizes a wide variety of awareness-raising events — plays, concerts, drawing competitions in schools — each dealing in its own often highly imaginative way with the risk of storms and the action one can take to reduce vulnerability. Thus, in a traditional puppet play, only the wise frog strengthens his house, while various frivolous farmyard creatures gossip and play around.[6] When the storm strikes the foolish creatures are left with nothing and, suitably chastened, listen to the frog's counsel.

Making the Message Sustainable

Families and official commune representatives must both actively participate if project activities are to continue after external funding comes to an end. DW therefore encourages the development of family groups. We invite beneficiary families to form a group in each targeted hamlet. The project works with the group to show how existing buildings can be strengthened simply and efficiently in a manner that is sympathetic to local tastes. Families always contribute both in kind and financially to the cost of strengthening. They are also involved in decisions about which families should have priority for support. These families then become a focal point for sharing information and experience with their neighbors.

At the commune level DW collaborates with the People's Committees to establish a Commune Damage Prevention Committee (CDPC) that progressively takes on responsibility for managing most of the project activities. The CDPC brings together members of the People's Committee, village representatives, and local unions. DW provides training and work sessions and helps the CDPC to organize activities in the commune. Once established the CDPC becomes the commune level extension of the project team, and in time takes over its activities.

Showing How the Message Works

Success requires three complementary, equally important, and interrelated actions: training, demonstration, and finance initiatives. Training workshops cover issues of typhoon damage prevention, practical theories and methods that are suited to the locality, and hands-on work on local buildings that demonstrates different techniques. Each participant receives a manual on cyclone resistant construction but is also encouraged to remember that each commune has its own styles and requirements and that few buildings are actually the same.

DEVELOPMENT WORKSHOP

DW works closely with individual families to demonstrate strengthening. Each beneficiary house is surveyed, its weaknesses discussed, and a contract drawn up detailing the work to be done to make the building safe. The contract also defines the contributions the family and the project will make. The project makes a subsidized contribution in the order of about US$150 and (since the beginning of 2002) the family has access to a credit fund to contribute to its share of the costs.

6-10: *Typical strengthening strategies include attaching roof elements together, holding down roof coverings, providing shutters and doors, and tying the structure together.*

DW also helps communes to strengthen small public buildings including primary schools, kindergartens, and markets. Work on these buildings provides additional exposure for techniques that are used on homes.

No two buildings have the same strengthening needs, but typical actions include ensuring that all parts of the structure are solidly tied together, that the roof covering is securely held down with reinforced ribs or bars, that the building has strong doors and shutters, and that walls are made more water and wind resistant.

Not only are people very ready to participate in the process of strengthening, but they will also borrow money at high interest to cover their contribution. DW found that this was causing hardship and set up a revolving credit scheme that grants modest loans to family groups. Families borrow amounts in the order of US$100 to US$150 and make regular repayments at 3 percent interest. The credit system is managed by the Commune Damage Prevention Committee in collaboration with the Women's or Farmers' Union.

Conclusion

In 1999, when DW first proposed the project, many people were sceptical that it could work or even be of value. Four years later opinions are very different. DW's team in Thua Thien Hué has been successful in changing the attitudes and the practices of families, technicians, and decision makers, and damage prevention in housing has moved much higher up the agenda for action in a country still facing many challenges. The challenge now is to bring more help, both practically and financially, to vulnerable families so that they, too, can "vaccinate" their homes and public buildings against storm damage.

DEVELOPMENT WORKSHOP

6-11: *Beneficiary families are selected by villagers and the commune according to social need. Development Workshop encourages a process that comes from the bottom up and assists family groups to work together to manage credit and help.*

Case Study:
Normal Life after Disasters?
Eight Years of Housing Lessons from Marathwada to Gujarat

Alex Salazar

In housing disasters, what gets built afterward is often created out of heated political passions, designers' egos, and the financial needs of developers and banking institutions — with only lip service being given to peoples' participation. Too often, unfortunately, this has led to the mass relocation of communities. Over the last 30 years, probably tens of thousands of villages and towns have been relocated to make way for development work (dams, aqueducts, frontier development, post-disaster rehabilitation, etc.), with many of these projects becoming well-known failures: abandoned or never occupied as people return to and rebuild at their old settlement sites.

After the 1993 earthquake in India in the core disaster-affected area of Marathwada, donor and World Bank/Government finance projects followed much the same pattern. Take, for instance, the story of Sugreev — a resident of new Gubal village, where about 225 families were relocated into geodesic domes. Unlike upper income families who were able to demand rectangular homes from the government, her family had no choice but to accept the donor organization's social experiment of geodesic living. Although lightweight and seismically sound, the dome form has proven to be completely inadequate for anything other than storage, and even for this it seems awkward to her.

ALEX SALAZAR

6-12: *Abandoned geodesic domes in Gubal. Most relief efforts are useless at best and destructive at worst. These structures, donated by an aid agency, do not reflect the traditional patterns of the local people and thus have been rejected by them.*

Nonetheless, like other marginal farmers in the village (who own, perhaps, a few acres of land and a cow), Sugreev's family has accepted the dome and coped with this purely technical solution to earthquake hazards by spending tens of thousands of additional rupees (not counting their labor) to build two additional rectangular rooms: one for sleeping and another for cooking. These are much more to her liking and have helped create a more traditional, rectangular wada-style courtyard house so typical to this region. To pay back the construction loan, both her son and husband are now indentured to a local landlord for two to four years at one-half their normal income.

While this example may seem extreme, it is not an uncommon tale. Indeed in the wake of the Marathwada earthquake, Sugreev's story typifies just one of many problems villagers face, as well as the overwhelming will of residents to make do with what they have, to live life in as normal a way as possible. Given the recent Gujarat earthquake disaster, it seems important to review some of the lessons learned over the last eight years in Marathwada; and, by comparison, to assess current housing issues faced by government and nongovernmental organizations in Gujarat.

Home Sweet Homelessness

Although Gubal's domes may be at one extreme of inappropriate house design, the rectangular bungalows created by government and nongovernmental donors in Marathwada are only marginally better. In fact, in the 52 core disaster-affected villages eight years after the quake, only about half the relocated families are sleeping indoors! Residents still distrust the quality of construction and are scared of repeat disasters. This is a legitimate concern, according to local engineers who observed minimal curing of concrete at many construction sites (owing to drought) and a lack of oversight and enforcement of building codes. At one village it was widely reported that a villager demonstrated how shoddy the construction was by breaking a concrete block over his head!

On the other hand, some projects are supposedly fine in terms of technical construction. Yet villagers find enclosed, interior rooms uncomfortable. As with the residents at Gubal, most use enclosed spaces only for storage, preferring to build their own temporary (kutcha) and permanent (pucca) extensions for day-to-day living. For the few middle and upper income families who received Government of Maharashtra supplied housing, such incremental, self-help building was allowed for. Thus at places like Utka village, one can find new houses in various stages of expansion, some with house plots already completely built-out, closely following the traditional wada-style courtyard house form.

Yet even at government-designed villages, many problems remain:

- No effective earthquake education/construction training has occurred in the core, relocated villages. While house extensions are being built, there is little connection between old and new walls, and new masonry work repeats many of the same basic masonry mistakes as in the past. Villagers continue to live in extremely dangerous housing conditions, despite all the time and capital spent to relocate them.
- Bathrooms, which have been provided at each housing unit, are not used. Villagers have preferred to relieve themselves in the fields, in the streets, or

in the corner of their compound, as they normally did in the past. Thus nearly all bathrooms are used only for storage.

- Showers, which were also provided at each housing unit, are being used. However, nearly all families have redesigned them in the traditional vattal style, with built-in water storage and heating (burning scrap wood or cow dung). This indicates that villagers would have built their own shower facilities had it been left to a self-help construction process.
- Roof leaks and excessive light and air infiltration through large windows are occasional complaints by villagers who find their new homes uncomfortable. Generally, however, they have simply learned to cope with these conditions — any shelter being better than none. At some places where leaky roofs have been a major problem, donors or the government have returned to do patch-up work.
- Compound walls, for most families, are still unaffordable and only partially complete. This has left residents with no privacy and exposed them to theft and the harsh local climate. The lack of compound walls has had a similar negative impact on public open spaces and streets.

Planning for Relocation and Abandonment

At a village planning level, relocated families are also living under unnecessarily difficult housing conditions. Although trees are beginning to take root and provide shade to the sprawling streetscapes, the reorganization of village life into industrial townships, the rigidity of planning methods, the immense scale of new villages, and the lack of sustained maintenance of public infrastructure and buildings have undermined the quality of the built environment.

Some obvious problems are:

- Local village councils (panchayats) have not had the financial and/or technical capacity to maintain excessively sprawling public infrastructure.
- Practically all storm water drainage systems are in complete disrepair, silted up, and unmaintained.
- One-half to two-thirds of road widths are not used. They were designed excessively wide.
- New centralized shopping complexes that open onto common public spaces are mostly deserted. Shop owners have preferred to build shops along the road in front of their new homes, just as many did at old villages.

Perhaps most problematic for villagers has been the uneven, often random reorganization of social relations owing to hasty planning decisions. In some places where villagers have remained in their communal groups (but are now located far from each other due to the immense scale of the new village), caste-related problems have reportedly become exacerbated. In other places where the mixing of community and income groups occurred, there are reportedly improved social relations, but some complain that they now live far from friends and relatives.

Complicating this is the widespread problem of living too far from one's farmland. The scale of new villages (5 to 10 times as large as old villages) has translated into less free time and an increase in household transportation costs, mostly through motorcycle and jeep fares, which are particularly taxing on the poor. This has led a number of families to shift into self-built homes on their farmlands. Not surprisingly, some wealthy families (who own but do not farm the land) have even shifted into second homes in small towns, preferring to rent out or abandon their relocated homes all together. Not even in new Killari village, where extensive institutional and professional design work was done, has this problem been resolved: entire rows of houses can be found vacant eight years after the quake!

From Marathwada to Gujarat: Reconstruction In-situ and Retrofitting

Relocation projects in Marathwada — both in terms of house design and settlement planning — were very poorly done, with few exceptions. This is a reality well known to housing professionals in India but unfortunately ignored by some government, World Bank, and US research organizations. Nonetheless many of the same Indian agencies and individuals involved in Marathwada are also involved in Gujarat, and their experience is invaluable. While some relocation work is going on in the same old, wrong-headed way, there also is a lot of momentum to repair and rebuild settlements in-situ. This general shift in policy and practice is the outcome of several factors.

First the social and physical geography of Gujarat is extremely diverse and difficult for organizations to handle with any one general approach. The size of Kutch alone, the main district affected by the quake, is around four times as large as the core area affected in the 1993 Marathwada quake. The settlements in Kutch are also generally much farther apart, making transportation costs, staffing, and material production much more costly. And the number and diversity of populations affected — which ranges from desert goatherding families in small hamlets of 3 or 4 homes, to shopkeepers in large urban settlements — has necessitated a decentralized approach.

Second it has been widely reported that people in Kutch are familiar with disasters, especially repetitive cyclones. With the lack of government aid in the past, people have become accustomed to rebuilding in-situ on their own. Thus, after the quake, the initial call for relocation did not go over well; people did not trust that aid would ever come through, and most did not see the need to relocate.

Government policymakers have begun to recognize these complexities, as well the problems with relocation work in Marathwada, and have subsequently crafted polices that leaves the decision to relocate to local residents. Government has also taken the position that it will generally only supply funding for individual homeowners, and it will avoid being in the business of building new homes and settlements. This has allowed new house construction in the region to be done in a more decentralized manner, with most households rebuilding homes in ways they are accustomed to, using skilled and unskilled labor hired independently or secured through NGO/donor involvement. Most of this work is in-situ reconstruction.

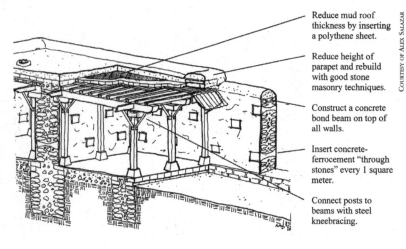

Reduce mud roof thickness by inserting a polythene sheet.

Reduce height of parapet and rebuild with good stone masonry techniques.

Construct a concrete bond beam on top of all walls.

Insert concrete-ferrocement "through stones" every 1 square meter.

Connect posts to beams with steel kneebracing.

COURTESY OF ALEX SALAZAR

This approach, although an improvement over work done in Marathwada, is not without problems. There are now a variety of housing solutions being created with little regulation and control on building standards. Many of these repeat the same basic masonry problems as before. Moreover the relatively few relocation projects being carried out are of varying degrees of quality.

These problems overlap with what are now critical issues: technology transfer and retrofitting. In Marathwada two basic efforts were used after the 1993 quake. One partially successful approach was initially engineered

6-13: Simple strategies for retrofitting traditional architecture for better performance during a natural disaster. Getting this knowledge to the people and having it used has proven to be a major challenge.

by Dr. Aria (Roorkie University, UP), and then refined in the field by Rajendra and Rupal Desai of the National Centre for Peoples' Action In Disaster Preparedness (CEDAP, Ahmedabad), who worked with a team of architects and engineers from Ahmedabad Study Action Group (ASAG, Ahmedabad). The techniques used consisted of several basic components to improve the seismic performance of traditional wada-style vernacular buildings. Core to this strategy was to work directly with homeowners and local artisans, thus transferring the technology into the local building culture. Components included: reducing the weight of heavy mud roofs by introducing plastic sheets for water protection; adding steel "knee" bracing to strengthen timber substructures; adding reinforced concrete beams at the top of existing walls; inserting concrete "through stones" in existing stone and mud walls; and several other components. To this day neighbors ask the local beneficiaries for their advice on how to build seismically sound vernacular buildings. However although successful at an individual level, the methodology has not caught on at a larger scale.

A parallel attempt to popularize retrofitting was attempted in the official World Bank and Government of Maharashtra program, which relied on the creation of women's groups in thousands of villages. Although the creation of women's groups may have been beneficial for equity issues, the program ended up promoting the construction of new room additions using brick masonry. Thus, while each family may indeed have one new earthquake safe room, villagers throughout the periphery are still living in damaged stone masonry buildings and have little understanding of proper stone masonry construction.

As any casual visit to Marathwada will reveal, villagers and artisans continue to build in both brick and stone, and they are repeating the same basic masonry mistakes as before. And so we arrive at a major dilemma for the rebuilding effort in Gujarat: how to learn from the technology transfer and retrofitting experiments in Marathwada while insuring that homeowners and artisans actually end up using these techniques in the future. Fortunately there is an ongoing project through the Gujarat State Disaster Management Authority (GSDMA) to train government engineers in earthquake safe building and retrofitting methods under the guidance of Dr. Aria and the CEDAP team. The government is also moving toward allowing a second installment of financial aid to be used by homeowners to retrofit work completed with first installment funds. Only time will tell if this new effort will work in the field, and if disaster-safe building techniques will be absorbed into the local building culture.

Housing Experiments: Some Final Thoughts

Perhaps the most encouraging news from Gujarat is the innovative, collaborative effort between NGOs, the United Nations Development Programme (UNDP), and the government

ALEX SALAZAR

6-14: *Typical earthen-walled and thatched houses in India. Without adequate structural reinforcement, such houses are susceptible to natural disasters. However, their replacement with "modern" alternatives after such events is often a disaster in itself.*

of Gujarat. Under the Kutch Navnirman Abhiyan NGO umbrella, a group of about 20 NGOs are receiving technical guidance in building technologies, planning, and house designs for their work at a grassroots level in about 200 villages. UNDP is serving as an intermediate link between NGOs and the government — both for issuing policy and for integrating feedback from the grassroots into policy decisions. At a local level the results seem very promising, with in-situ and relocation work being done using low-cost technologies, mostly local materials, and an eye toward vernacular planning and house design patterns. While there are bound to be difficulties with this effort, this is a new, innovative experiment in social-political organization after disasters. And it is a major step forward out of the political morass and NGO infighting that characterized the work in Marathwada. The shift in policy and practice speaks volumes about the will of various organizations to embrace a self-help ethic that respects the need of disaster-affected communities to return to living as normal a life as possible.

The author would like to thank the Graham Foundation (Chicago) for their generous 1993 grant that supported the initial research of this paper. The author would also like to thank Architect Amol Gowande (Latur) and apprentice Architect Venkatesh Thota for their assistance and contributions to research done in 2001.

CHAPTER 7

A Critical Overview of
Sustainable Building Techniques

Joseph F. Kennedy

A critical overview of the range of basic building materials and methods available to the majority of the world's population can help builders make wise choices. This chapter is a brief attempt to illustrate the pros and cons of various systems from a technological, cultural, and ecological standpoint. The goal is to create a new, sustainable architecture with "hybrid vigor" that has as its roots vernacular tradition judiciously grafted with industrialized methods.

Foundations

A foundation is the first part of a building to be constructed, and the most important for longevity. A foundation must help the building resist gravity, wind, and earthquakes as well as stop it from sliding downhill. There are three basic purposes for a foundation:

- It spreads the weight of the walls and roof — especially for "point loads" (concentrated weight caused by posts, etc.)
- It holds the building together as one unit by providing continuous reinforcement.
- It raises walls off the ground and protects them from water

Stone Foundations

A stone foundation is built using stones that are either mortared (with mud [weaker] or concrete) or dry-stacked (no mortar) to create a surface for the wall to rest upon. (Angular stones are more stable than round ones.) It is important to lay the stones "one over two" so as not to have continuous joints, and to use stabilizing "tie stones" occasionally through the width of the wall. Building with stone can by a time-consuming, highly skilled craft that is even more difficult if appropriate stone is not available. Building with

KELLY LERNER

7-1: *Stone remains an excellent foundation material, available in many places in the world. Adequate reinforcement is critical, especially in areas subject to earthquakes or other natural disasters.*

forms and concrete mortar (see "Teaching Sustainable Settlement Design in Lesotho") can help speed up the process. To increase strength in stone foundations, especially in seismic areas, a "bond beam" of concrete with steel or other reinforcement can be created on top.

Concrete

Making concrete is fairly easy to learn, but it is susceptible to shoddy building practices (see "Capacity Building on the Periphery of Brasilia, Brazil"). Concrete is made of lime, sand, and Portland cement mixed with water which, when cured, doesn't break down for many years (if made correctly). Because concrete is strong only in compression, reinforcement (usually of steel) is often added for tensile strength. Some efforts have been made to replace steel with natural tensile members such as bamboo. Producing the cement that goes into concrete is a highly polluting and energy-intensive process, and the resulting product can be prohibitively expensive in cash-poor regions. Some of the cement can be replaced with fly ash or rice hull ash to decrease cost and increase strength. By adding stones to the concrete as it is poured, less concrete may be needed. Some natural builders are using recycled concrete to build foundations and walls, but in those cases a poured concrete bond beam would be desirable on top.

Rubble Trench Foundation

A rubble trench foundation is a trench filled with large gravel (though small gravel or even sand could be used) with a drain to daylight or a dry well to take away water. A concrete grade beam is usually poured on top of the gravel to lift the wall above the ground and to provide continuous reinforcement. This type of foundation can greatly reduce the amount of concrete needed. It depends on a good source of gravel, the mining or making of which can be an ecologically destructive activity.

Other Foundations

Wood or bamboo poles sunk into the ground can be used as a foundation. This method raises the house up into the air, providing airflow underneath: in some climates (wet and humid) this can be an advantage. The poles have a tendency to rot rather quickly, however, which can be extremely difficult to fix. The poles can be placed on top of concrete or stone pier blocks for longevity, but this necessitates additional structural members and hardware for stability.

Another experimental foundation uses old tires filled with soil stacked in a running bond, like bricks. This system uses a waste product, but it is very labor-intensive and its health effects are still being debated.

Woven polypropylene bags filled with soil or gravel, which are then compressed with a tamper, are another foundation option. Bags are cheap and easily transported, but they can break down in the sun if not plastered. Soil-filled tires and earthbags can be used to create entire walls (see "Walls," that follows, for a more detailed description).

Walls

Many different wall systems can be used, some in combination with others. The choice of wall system(s) depends on many different factors. These include:

- Esthetics
- Availability of materials
- Climate (sun, humidity, rain, snow)
- Ease of working
- Cultural factors (what is traditionally used in the area)
- Site conditions (terrain, shade/sun, fire danger, etc.)
- Privacy needs/neighbors/noise
- Cost
- Whether or note the building is in an urban or rural building environment.

The following describes some common wall systems.

Adobe

Adobe construction uses sun-dried mud bricks (adobes) stacked with a mud mortar to create thick walls, and even curved roofs (only recommended in the driest climates). Like all earthbuilding techniques, adobe is an excellent source of thermal mass but may need additional insulation in colder areas. While adobe is weak in tension, it is strong in compression.

Adobe bricks are made with a mixture of clay and sand (and sometimes straw or manure) combined with water and then poured or pressed into forms that are later removed. Asphalt emulsion is sometimes added to the mix to repel moisture, especially for lower courses of adobes. Adobes are laid on an appropriate foundation (usually stone or concrete) to avoid moisture damage. As adobes are irregular in size and shape, thick mortar joints are needed to make up the difference in size of the adobes, adding significantly to the skill needed to build a straight and plumb wall.

Wide eaves are often necessary to protect the walls from rain, and foundations must be high enough to protect the walls from splashing and ground moisture. Mud plaster is traditionally used to finish adobe structures, requiring renewal every few years. Adobe construction is particularly low-tech and needs a minimum of tools. Making the bricks can be a good micro-enterprise. Because the bricks form "cold joints" at the mortar/brick connection, the system is more susceptible to earthquakes than are monolithic systems such as rammed earth or cob. Continuous grade beams and bond beams can help adobe survive earthquakes better (see "Low-Cost Housing Projects Using Earth, Sand, and Bamboo").

Cob

Cob is an ancient technique for building monolithic walls using "cobs" of moist earth and straw and is one of the best earth-building techniques for use in seismic areas. Cob is easy to work with, has wonderful sculptural qualities, requires few tools, and can be made by young and old alike.

To create cob, one mixes local soil with sand and/or clay (depending on the composition of the soil) and straw or other fibers to create a stiff mud that is formed into small loaves (cobs). Cobs are then tossed to a builder on the wall, who mashes them together to form the wall on top of a stone or concrete foundation. Relatively thick walls are built in layers from 6 to 18 inches (15 to 46 centimeters) high (attempts to go higher can result in slumping). After a period of time to let the layer solidify, work can continue. Irregularities are shaved off with a spade or other sharp tool as work progresses. Windows and other details can be cobbed into place, and niches and reliefs are easy to create. Cob is also useful in combination with other techniques such as straw bale construction.

Although cob can be extremely inexpensive, it takes a lot of time and effort to make. Some builders are investigating ways to use commonly available machinery to speed up the cob-making process. Others are experimenting with making cob in forms to speed up the building process and to get straighter walls.

Compressed Earth Blocks

Compressed earth blocks are similar to adobes, but they use less water, are denser, and much more uniform in shape. Blocks (which are sometimes stabilized with cement) are created using a variety of machines. Some, such as the Cinva-Ram (invented in South America), use human labor and are relatively inexpensive. Others produce thousands of bricks in a day but use expensive fuel-powered machines. Either option provides a good opportunity for micro-enterprise. Because of their uniformity compressed earth blocks need little or no mortar, can be laid up rapidly, and produce straight, precise walls. Compressed earth blocks are a good example of how natural materials can be used to create a "modern" look.

Earthbags

Earthbags are soil-filled fabric sacks or tubes that can be used to create foundations, walls, and domes. Moistened soil (which is sometimes stabilized with cement) is stuffed into a bag or tube set on the wall and then lowered into place. The filled bag is then compressed using a hand tamper. In earthquake-prone areas, a layer of barbed wire can be used as "mortar" between the bags to prevent slipping.

Clay-rich soil can be used with weaker burlap bags (the compressed soil becomes strong enough to hold itself up, allowing the bag material to be cut away from the surface). Stronger, structural polypropylene bags are preferable for sandy soils. Long tubes of the bag material can be filled and stacked like a coiled ceramic pot. Recycled sacks are often available free or at little cost. Because plaster sometimes does not stick well to the bags, reinforcing mesh is often advised, which can be attached to the wall by strings laid between the courses of bags.

Many natural builders use earthbags as a simple foundation for straw bale or cob structures, or for simple site walls. Gravel-filled bags have been used as foundations to avoid capillary action. Although structural tests have been done on the system, it is still in its infancy, and long-term performance is as yet unknown.

Leichtlehm

Leichtlehm (literally, "light-clay") is a German technique in which loose straw, wood chips or other fibers are coated with a clay slurry and tamped into forms. With "straw-clay," the most common of the systems, straw and a clay slurry are tossed with pitchforks or mixed mechanically and allowed to age for a few hours. (This allows the straw to absorb extra moisture to create a stickier and more easily tamped mixture.) For more insulation, less clay or lighter tamping can be used. The straw-clay mixture is hand-tamped between forms in 2-foot (0.6-meter) layers. Once each layer is complete, the form is moved up and the next layer is tamped until the wall is complete. Walls are allowed to dry before plastering. Shrinkage is taken up by stuffing more of the mixture into the cracks. Because the load-bearing capacity of the material is variable, it is generally used as infill between supporting elements (timber frame or post and beam).

Straw-clay stuffed loose between rafters has been used as insulation (the clay discourages pests). It has also been used as an insulating layer underneath earth floors.

CATHERINE WANEK

7-2: *Straw and clay mixed together (either with a machine like that shown or by hand) make an insulative and attractive wall material. This technique may be difficult to use in areas without sufficient materials to make the necessary formwork for traditional straw-clay infill, though the development of straw-clay blocks shows much promise.*

Straw-clay tiles can be placed between roof rafters as insulation and as a plastering surface. Because of the formwork necessary for the wall system, however, it has been slow to gain acceptance in traditional areas. A system that uses straw-clay bricks as lightweight adobes, though, has been adopted much more readily (see "Straw-Clay Blocks").

Wood chips or other materials can be mixed with the clay instead of straw. These materials are poured into forms instead of tamped or are used to create lightweight bricks.

Rammed Earth

Rammed earth has excellent thermal mass (but needs insulation in cool climates), strength, comfort, and beauty. Rammed earth can be built with simple forms and tools and can be built in a variety of climates. Walls do not need to be plastered and will last for hundreds, even thousands of years (the Great Wall of China is partially built of rammed earth).

Rammed earth is built on a foundation of stone or concrete. A soil mixture of 20 percent clay with a moisture content of 10 percent is rammed using mechanical or hand tampers in layers or lifts of 6 to 8 inches (15 to 20 centimeters). The soil is sometimes stabilized with cement, and different soil types can be layered to create decorative effects. The wall is usually topped by a bond beam that ties together the walls and supports the roof. Rammed earth is another system that can be used to create a modern style of architecture, but the complex formwork necessary for precise construction can be daunting to unskilled builders.

Straw Bale Construction

Straw bales are bricks of straw created by compressing and tying loose straw using a baling machine. Although most common in North America, bale buildings have been built around the world. Originally used by the pioneers of the Nebraska sandhills, straw bales are cheap to buy and easy to build with.

Straw is an annually renewable material, available wherever grain crops are grown. It is often a waste product, much of which is currently burned in the field. Bales are easy to work with, lightweight, and require a minimum of tools. And the thick walls are good insulators. With a natural plaster, straw bale walls "breathe" (transpire water vapor), which together with their sound-absorbing qualities, provide a quiet, healthy interior environment. Straw bales can also be combined easily with other natural building systems.

Straw bales are commonly used as infill in a post and beam structure, or as a load-bearing system, where the bales themselves support the weight of the roof. Bales are attached to a concrete, stone, or earthbag foundation with pins or strapping. They are laid like bricks and externally pinned together using rebar, wood stakes, or bamboo (see "Casas Que Cantan"). A wood or concrete bond beam is laid on top of the straw bale wall, and the roof attached to the bond beam. The bales can be plastered with mud, lime, or cement plaster. In many cases stucco netting is not needed, and plaster can be applied directly to the bales.

"Construction grade" bales may be difficult to obtain in some areas, and the bales can be susceptible to moisture damage in areas without access to tarps, especially in load-bearing systems. Although they are quick to build with, details are time consuming to finish. Some cultures have had difficulty accepting bales as a valid building material.

Wattle and Daub

The technique of weaving branches (wattle) as a support for mud or lime plaster (daub) is perhaps the oldest of earth-building techniques and is still used for traditional architecture in many parts of the world. This system is particularly suited for thin walls in tropical areas and for seismic zones.

Wood

Wood is an ideal building material: strong in compression and tension, easily worked, and beautiful. Its major disadvantages are that it burns easily, and that its overuse is leading to the widespread destruction of forests. Building with exposed frames of wood surrounded by materials such as straw-clay or straw bales can take advantage of the beauty and structure of wood while using less of it. Using alternative materials such as cob or straw bales can lessen the need for wood and reduce pressure on damaged forest ecosystems. Some builders are finding uses for driftwood and irregularly shaped trees that would otherwise go to waste. Wood scraps and sawdust can be used for panels, hybrid products, etc., though many of these use toxic glues.

The wise use of wood is closely tied to sustainable forestry practices. Careful harvesting of trees can provide necessary materials while saving delicate ecosystems, and the use of smaller diameter trees or unmilled lumber can save large old-growth forests (see "Small Diameter Wood: An Underused Building Material").

Cordwood

Cordwood is a traditional technique in Northeast America and other heavily forested areas. Small lengths of wood held together with a cement- or mud-based mortar create thick walls. An insulation layer of lime-treated sawdust is used between the inner and outer mortar layers to keep the house warm.

JOSEPH F. KENNEDY

7-3: Wattle and daub uses small-diameter wood plastered with earth or lime plaster. It is an excellent infill system for moderate and benign climates.

Cordwood can take advantage of plentiful small-diameter trees not useful for other building purposes. It is relatively quick and easy to build with. A disadvantage is that differing rates of contraction can result in small gaps between the mortar and wood, leading to air or insect infiltration.

JON SOJKOWSKI

Concrete

Concrete, in the form of poured concrete or concrete blocks, has quickly gained favor as a building material for walls around the world. It is a prestige material, often used for status in low-income areas despite its often-inferior thermal, ecological, and cost performance. Concrete, if built correctly with sufficient tensile reinforcement, has well-documented structural characteristics. But it is often built with inferior ingredients or with improper methods, which diminishes its strength enormously.

Concrete can be appropriate in hybrid systems that also use other materials. It can be used to make structural posts that can then be infilled with other materials. It is also good for bond beams on top of earthen or straw bale walls.

7-4: Bricks are a desirable building material throughout the world. Most brick kilns, however, are extremely inefficient and are a major contributor to deforestation and pollution. Efforts have been made by numerous groups to increase kiln efficiency in order to address these drawbacks.

Fired Bricks

Fired clay bricks are a popular building material around the world. They lend themselves to local manufacture, are easy to build with, and are attractive and relatively weather resistant. Bricks are mortared together, usually with a lime- or cement-based mortar and can be used to create domed or vaulted roofs. The manufacture of bricks, however, has proven to be extremely ecologically destructive, both in mining the relatively pure clay necessary for good bricks, but especially in the highly polluting inefficient kilns commonly used to fire them (see "Capacity Building on the Periphery of Brasilia, Brazil"). Efforts to improve kiln efficiency may make this material more sustainable.

Hybrid Structures

The hybrid building concept combines several construction techniques to create a better building overall. For example the combination of a thermal-mass technique, such as cob or rammed earth on the sunny side of a house along with an insulating system, such as straw bales or straw-clay on the shady side, takes advantage of the best qualities of each. Hybrid structures may prove to be the best strategy for making highly efficient, desirable buildings with local materials.

Roofs

To protect against the elements, a roof must be sturdy and waterproof. A structural system (bond beam or top plate) is often built on top of the wall to support the roof itself. A layer of sheathing is usually placed on top of the structure, then a layer of waterproofing, and finally the protective roofing material itself. Walls must be designed to take the weight of the roof and of any loads that might be on the roof. These loads include snow, wind (uplift), horizontal forces, and people.

The greater the weather a roof must withstand, the steeper it is usually built. This is why you see steep roofs in places with lots of snow, and shallow roofs in milder climates. Roofs have to:

- Protect against rain, snow, wind, and excess light. This means they need to be solid, strong, and waterproof.
- Prevent unwanted heat loss or gain. As heat rises most heat loss occurs through the roof; they must therefore be well insulated. As roofs are also exposed to the sun, they must keep out the sun's heat through insulation or radiant barriers.
- Remain in place in hurricanes, earthquakes, or high winds. This means the roof must be well constructed and well attached to the walls and foundation. Straps or metal hardware are usually used to connect the roof to the walls. A bond beam is very important to create a continuous strong surface to support the roof load. Another strategy is to support the roof on flexible members such as poles, while the mass walls are built separately. This allows the two systems to move independently during an earthquake. (See "Low-Cost Housing Projects Using Earth, Sand, and Bamboo.")
- Resist extra snow and other weight (such as people on the roof). A roof must be designed to avoid collapse in a worst-case scenario.
- Shed water away from wall. This is achieved by an overhang, and the greater the danger to the walls from weather, the bigger the overhangs should be. Overhangs should also be designed correctly to block the summer sun and allow the winter sun in.

Roof Materials

Bamboo

Bamboo is the largest of the grass family of plants. It grows very quickly, providing renewable material for building, tools, and utensils as well as edible shoots. Common in the tropics, many species of bamboo grow in temperate climates as well. Bamboo is particularly suitable for creating beautiful roof structures and is extremely resistant to earthquakes. Bamboo can replace wood and steel in many other situations as well. It can replace rebar, act as pins in straw bale construction, create trusses and other structural members, provide decorative elements, and even function as plumbing. Bamboo is susceptible to insect damage, however, and must be treated for longevity. Sustainable forestry practices must also protect against over-harvesting. Although bamboo is considered a "poor person's material" in many countries where it has been used vernacularly, that association is changing as new structural systems allow for more ambitious architecture that is patronized by the wealthy (see "The Revival of Bamboo Construction in Colombia").

7-5: A traditional village in Japan demonstrates timber framing, wattle and daub infill, limewash, and thatched roofs.

Thatch

The use of reeds, grasses, or palm fronds as a roofing material is a traditional system still used in Europe and many traditional societies. It is one of the only entirely natural roof systems available, and many builders are exploring its use to replace methods that rely on manufactured materials. Thatched roofs, if well built, can last up to 60 years and provide a beautiful finishing touch to many natural wall systems. Thatch breathes, makes use of local materials, is highly insulating, and extremely beautiful.

Thatching is a highly skilled and time-consuming craft, with few skilled practitioners available. Hence a thatched roof can be very expensive. A poorly built thatch roof tends to leak and will need replacing within a few years. Thatched roofs must be protected from fire, and homeowners must be on the lookout for undesirable pests that find thatch an ideal home. In many countries, thatching materials are increasingly rare and expensive. (see the Text Box, "Why We Don't Build With Thatch").

Wood

Wood is still perhaps the most common roofing material, because it is easy to work and strong in tension (see "Wood" in the preceding section titled "Walls"). Trusses, which are manufactured structural components made of several pieces of small-dimension wood, can be used instead of ridge beams and rafters (see "The Pallet Truss"). Because trusses are very strong, they are often used to cover large spaces.

Living Roofs

Living roofs are beautiful, help the house blend into its environment, and provide climatic stabilization. They are not very well suited to dry climates because of the need to provide irrigation for most of the year.

A living roof is built on top of a sufficiently strong frame, with carefully applied waterproofing (it is very difficult to locate leaks once the sod or other growing material is in place). The living roof itself can be composed of a base of straw; once the straw decomposes, native or introduced plants can take root.

A living roof needs ongoing tending and can be a fire hazard in hot, dry climates. Its main advantages are beauty, its protective qualities (a living roof prevents damage to the waterproofing by ultraviolet radiation), and that it obviates the need for tiles or other shingles. Unless they are made of natural rubber, some membranes can be toxic, and any water collected off the roof should not be used for drinking.

Metal

Metal roofs have some very useful advantages: they are relatively inexpensive compared to other systems and are durable, easy to install, and lightweight (allowing the underlying roof structure to be lighter as well). Metal roofs can be used to collect water, a very important consideration in dry areas. Although metal takes a lot of energy to produce and the mining of metal ore is environmentally destructive, metal roofs are also easily recyclable. If uninsulated, metal roofs can make a room unbearably hot or cold.

Micro Concrete Tiles

Micro Concrete tiles are a relatively new roofing system that has gained much popularity in recent years, especially in Central and South America. Although the cement involved is relatively costly, the system lends itself to local enterprise, and produces an excellent product that is well received by a wide range of builders (see "Micro Concrete Roofing Tiles").

Shakes and Shingles

Shakes are overlapping pieces of wood used to shed water and snow. They are created by splitting a log into thin pieces using a hammer and froe. Shingles are similar, but they are created by sawing the log. Shakes and shingles used to be very common, but because old-growth trees are needed to make them, their use is no longer practical except in a few very specific circumstances.

Domes and Vaults

Domes and vaults are most common in the Middle East and are made with materials such as bricks, stones, and earth that are strong in compression. A vault is a roof that is curved in a single direction and is best for covering a rectangular shaped room. A dome is curved in two directions and can cover a square or round room. Combinations of vaults and domes can be infinitely varied and extremely beautiful.

Building with these systems is not common in many countries, and it demands skill and practice to produce structurally sound roofs. Indeed building domes or vaults of earth can be dangerous: it is difficult to waterproof an earthen roof, and if the roof gets soaked with water it can collapse. Domes and vaults are best used in the driest of climates such as deserts and even then, provision must be made to quickly move any rainwater off of the roof and away from the walls (see "Woodless Construction").

Other Roofing Materials

Other roofing materials include stone — especially flat stones like slate. Slate can be attached to the underlying structure with nails and are overlapped like shingles. Ceramic tiles are common in many countries and can be an easy and beautiful roof option. Both slate and tiles can be used to collect water. Both are extremely heavy, however, and need a very strong roof structure to hold them up. Cut-up old tires can be used like tiles but may preclude water collection off the roof. Asphalt shingles, although easy and convenient to use, are extremely toxic to dispose of and are therefore to be avoided. Concrete roofs are common in many places where wood is in short supply, and the same pros and cons regarding concrete's use in walls apply here.

Waterproofing

The waterproofing layer is perhaps the most challenging aspect of the roof's structure. Paper or cloth soaked in tar are most commonly used for waterproofing. Plastic and rubber are other options, though both rely on nonrenewable petroleum resources. Old-time builders used layers of bark or relied on the roofing material itself (thatch, tiles) to shed water. These traditional methods had much more tendency to leak, however, and current lifestyles demand a better solution.

Insulation

Insulation is used to keep excess heat or cold from entering the building through the walls, roof, and floor and is most important in the roof. Insulation must be kept dry or it can no longer do its job. It must also be protected from animals, as it is a preferred nesting area.

Wool

Wool makes an excellent insulation, especially in areas where sheep are common. Some people even recycle old woolen clothes as insulation. It is one of the only materials that can insulate well when wet. If you put the wool in plastic bags, it will prevent moths and mask the smell wool sometimes has.

Straw

Straw is a moderately effective roof insulator. Some people put baled straw or loose straw in their roofs (see "Straw Bale Construction in Anapra, Mexico"). If bales are used, then the cracks between the bales must be carefully stuffed to avoid heat loss. Some builders mix straw with clay and either make insulating tiles or place it loose in the roof, though this does not usually provide sufficient insulation. Straw insulation can be a fire hazard. Clay slip applied to the bales has been used to prevent fire. Bales can be very heavy, as well and if used, the structure holding up the roof must be strong enough to support the weight.

Cellulose

Cellulose insulation is a popular option, as it is an inexpensive nontoxic recycled product (from old newspapers). Cellulose insulation is usually blown into the wall or roof cavity with a special machine.

Perlite/Vermiculite/Scoria

Perlite and vermiculite are porous minerals that have very good insulating qualities. They can be more expensive than other options but, if locally available, are particularly good in situations that might be exposed to water or fire.

Other Insulation Options

A variety of manufactured insulation products is available, of varying toxicity and expense. Most would be beyond the economic reach of builders in many parts of the world.

Wall Protection Systems

Plasters

Plasters are spread onto a wall surface for protective and/or decorative reasons. Exterior and interior plasters have different requirements and are often made of different materials. Plasters provide:

- Protection from weather
- Control of dust
- Color and decoration
- Texture
- Durability of the wall surface.

7-6: *The cob earthbuilding technique has been used in this house in Thailand to great sculptural effect. Cob construction necessitates few tools and is easy to learn. It can be fairly time consuming, however, and the high-mass walls may not be appropriate for all climates.*

Plasters are characterized as "hard" (cement-based) or "soft" (earth, gypsum, and lime). Cement-based plasters are often easier to maintain in the short term, as they are easy to clean and relatively resistant to water. Plus skilled plasterers are relatively available. Cement plasters are hard and cold to the touch, however, and have poor acoustical properties. And cement plasters are difficult to nail into and repair. Cement is caustic to the skin, and must be worked relatively quickly. Cement plaster is often used to plaster earthen walls but, as it can crack and let water in that catastrophically erodes the wall, it is not recommended for this application (see "Speaking the Vernacular").

Soft plasters are pleasing to the touch, easy to repair, and have good acoustical properties. They are harder to clean than cement plasters and need more frequent maintenance and replacement. Mud plaster (clay, sand, and fiber) is particularly forgiving and easy to apply, though it needs the most frequent replacement. Gypsum plasters are hard and smooth but not weather resistant, so are most appropriate for interior surfaces. As gypsum sets very quickly, it needs more skill to apply than do other soft plasters. If choosing between earth and cement plasters, lime plaster is an excellent compromise in strength, durability, and flexibility. Further lime has the potential to be locally manufactured (as opposed to cement, which is usually made in centralized factories). Lime plastering requires the greatest level of skill, however, and careful materials preparation.

Other Wall-Covering Options

Other wall-covering options include shingles, wood boards, gypsum board (interior), tiles, and "living walls." All these options have pros and cons, as previously discussed.

Paints, Sealants and Finishes

Most wall-covering systems include a final step of painting or sealing for added durability and decoration. Most modern paints and finishes, however, are extremely toxic and inappropriate for many natural building materials. Because of this, efforts to reclaim vernacular traditions are expanding. Traditional paints include those based on lime, natural oils (usually linseed), casein (milk protein), or clay, often with additives to improve durability or appearance. Traditional paints take more skill to apply and may not be as easily cleaned as modern paints. They also cannot be applied over modern paints without special treatment of the walls beforehand. Casein paints must be used quickly once made, as they soon spoil.

Traditional paints have a more "lively" appearance and, as they can transpire moisture more easily, are more appropriate for earth-, straw-, and lime-based wall systems, where release of moisture is most important. Most traditional paints are extremely stable on interior walls, but may need more frequent replacement on exterior surfaces.

Floors

Floors are usually the final part of the house installed. Mass floors (concrete, earth, brick, or stone) can be an excellent source of thermal mass for passive solar designs (see "Sustainable Building As Appropriate Technology"). Floor Options include:

- Concrete slab (expensive but durable; can be covered with tiles; can have embedded thermal heating)
- Bricks on sand (beautiful and easy to repair)
- Soil cement (this can also be made into pavers)
- Stone on sand (difficult to clean, but beautiful and durable)
- Earth (can be poured or tamped; is beautiful; is soft and warm to the touch; must be sealed with linseed oil or other hardener for durability; can have embedded thermal heating; can take a long time to dry)
- Wood (best for above-ground floors; potential for rotting if too near ground).

Conclusion

The building systems described in this chapter are not exhaustive but represent a wide selection of methods used throughout the world to build basic housing. As nearly all buildings are hybrids of different systems, specific combinations of these techniques need to be tested in the field for technological performance, cultural acceptance, and ecological sustainability. Although other issues such as land tenure, ecological destruction, and political crises are major factors affecting housing issues, a fully engaged set of building options can address the lack of housing around the world.

Case Study:
Affordable Housing from Local Materials in Ecuador

Darrel DeBoer

The housing shortage in Guayaquil, Ecuador is graphic and overwhelming. People spill in from the countryside seeking jobs, and the civil war in neighboring Colombia causes thousands of refugees to seek asylum. The result is hundreds of thousands of people, with family incomes under US$50 a month, crowded together and forced to survive with few to no resources. The rules of the streets of Guayaquil are not much different than in any frontier town, with vigilante justice and incredible violence over the possession of inconsequential items.

For the extremely poor (thought to be nearly a third of the Guayaquil's population), the only way to obtain land is through a "land invasion" — an organized collective that takes possession of government-owned land simply by their massive numbers. A leader coordinates a "takeover day," and thousands of people converge and build enough ad hoc buildings that the government can't kick them out. Then the leader uses connections with local officials to ensure the transition to a legal community. Over weeks and months, as the proper wheels are greased, true settlement begins. And as the community is built, self-governance takes over. No government services are available, so collective kitchens, schools, daycare, roads, police, and utilities are all invented.

The problem of homelessness often feels insurmountable, yet a workable solution is in process. A Catholic group, Hogar de Cristo (House of Christ), decided their work was inconsequential unless they could help fulfill people's fundamental right to safe housing. To that end they build 40 houses a day from inexpensive local materials. The least expensive house is US$360, with a US$10 delivery charge.

Jesuit-run, the Hogar de Cristo factory has become quite an institution. A thorough screening process requires a friend or neighbor to cosign for each house, and an interest free payment plan allows a family to pay off their house in three years with payments of just a few dollars a month. It is nearly impossible for North Americans to imagine how such an acute housing shortage can be solved without the use of public funds.

Sometimes professionals, such as architecture professor Jorge Moran, volunteer their services. Jorge Moran is one of the brightest stars in bamboo construction and has written numerous books documenting indigenous building techniques and forms. Living in the huge city of Guayaquil (with almost nothing to recommend it to outsiders), Jorge sees not the poverty but the people. His work with and advocacy for the Hogar de Christo housing project is the living example of his enthusiasm; his pride in what the

Jesuits have accomplished is infectious. Although the solution has its roots in desperation, the work is elegant and clear: there is a need for houses that people can deliver on a single truck, assemble in half a day, and genuinely afford to pay for so they can get back to the difficulties of daily life without missing a beat. And in this place of dire need, virtually everyone in the barrio greets him warmly, each congratulating the other on how gordo (affectionately, "fat") they have become.

Houses from Hogar de Cristo are only sold once proof of land title is shown, and the organization interviews prospective homeowners to determine who will benefit. Women are solely eligible to hold title (only women have been found responsible enough to continue to make payments), and those supporting families are given preference. As applicants progress through the interview process, they receive differently colored poker chips at each stage. This ingenious technique ensures that they can pick up where they left off on their last visit (it often takes three to five trips to provide all the information required). Cosigners are not allowed to be in the military or in any form of policing agency, nor are they allowed to be lawyers! Any person with a cell phone or beeper is not eligible. These requirements offer some insight into the social structure of Guayaquil, where a drug dealer/extortionist/vigilante culture is the way to short-term profits.

7-7: A view of Hogar de Cristo's factory in Guayaquil, Ecuador.

The houses are all built in a factory on the edge of Guayaquil. Designers have taken the archetypal, basic house and made it structurally sound and amazingly simple. Factory workers are paid by the piece and move faster than anyone else I've seen in Latin America. Workers align mangrove wood (generally very hard and straight) over a steel jig to form the house's frame. (Only the Jesuit builders of Hogar de Cristo are allowed to harvest the mangrove's aerial roots — overharvesting has led to banning others to access — and any wood confiscated from poachers is given to Hogar de Cristo.) Once the frame is ready, workers nail on split-open, flattened poles of bamboo as siding. The many gaps between the poles are large enough for much-needed ventilation, but small enough to provide security and some measure of rain protection. Galvanized corrugated metal roofing completes the basic kit. Since much of Guayaquil is in a flood zone, where water can be expected with every tide to rise at least one floor high, the footings and bases of the

7-8: *The components of a house ready for delivery.*

7-9: *An interior view of one of the houses by Hogar de Cristo, showing the lightweight frame, bamboo exterior panels, and metal roof.*

houses vary widely and are left up to the occupants to design and build ahead of time. The entire house can be stacked on top of a single truck, with assembly expected to be half a day's work by unskilled but relatively strong workers. Several hundred people visit the factory each day, hoping to buy a house.

The program is not without its critics. There is no plumbing, and electricity is illegally patched into local utility lines. Houses look good for only half a dozen years, due to a lack of anti-insect treatment for the bamboo. And there is as yet little concern for many of the preoccupations of the natural building movement in the US, such as reducing the use of cement and toxic materials. In Guayaquil, as in many parts of the world, health considerations fall way below survival worries — even Hogar de Cristo suggests that owners apply used diesel oil as an insecticidal remedy for bamboo.

On the positive side, although the houses do not have shear walls to resist earthquakes, they weigh almost nothing and are such an improvement over what people have that they certainly perform well enough. None of the members are large enough to do much damage if they land on you, and everything is repairable. Although temporary (with an expected life of a decade), with care many of these houses have lasted much longer. And their thermal performance is better than the concrete houses with which they are often replaced — the thin walls allow ventilation but do not provide mass to build up the sweltering Equatorial heat.

The affordable housing of so many people in Guayaquil is made possible by creative economic strategies and the use of appropriate local materials whose characteristics are known to local workers. The example of many effectively detailed structures allows the next generation of self-built structures to be that much more secure than they might have been. The bamboo houses of the Hogar de Cristo give owners fewer worries, and demand a lower percentage of household income than that faced by new homeowners in the US. In Guayaquil, owning a home in a place with such economic insecurity allows the occupants to focus on other challenges and gradually create a decent way of life.

As family fortunes improve, will permanent bamboo houses replace the temporary houses of Guayaquil? Time will tell. In neighboring Colombia, environmentally safe insect/fungal treatments and improved earthquake performance have convinced the wealthy to accept bamboo as a standard structural material. But in Ecuador the wealthy still set the example by building almost entirely in concrete. Yet even in Ecuador, a bamboo conference I attended attracted more than 800 enthusiastic participants who were looking for good information to improve their building methods. So perhaps Ecuadorians, too, will eventually be won over to the amazing local resource of bamboo.

THE REVIVAL OF BAMBOO CONSTRUCTION IN COLOMBIA

Darrel DeBoer

The traditional architecture of Colombia originally relied on the huge timber bamboo *Guadua Angustifolia* for most construction in urban and rural areas. After fires ravaged bamboo structures in several Colombian cities a hundred years ago, however, the remaining buildings were covered with plaster, and people subsequently built with brick and concrete. While it leaves much to be desired thermally and its performance can be catastrophic in severe earthquakes, concrete won over the culture during the 20th century. In recent years, however, a few pioneering Colombian architects have worked to revive bamboo construction and bring it to complete acceptance again.

Some of the finest buildings of any type I've ever seen have been made by Simón Vélez, Marcelo Villegas, and their workers. This group invented a joinery system and a viable system of preservation that has allowed bamboo structures to succeed in the modern, engineered world. Their methods do not transfer easily, however (the group doesn't do engineering calculations, believing that the current formulas don't adequately reflect what bamboo is capable of). Thus their initial use of bamboo (though structurally brilliant) had been confined to a few, highly experienced people.

With the invention of the bolted/filled joint, however, the groundwork was laid for the current renaissance in bamboo construction. Vélez and Villegas built some very large, impressive buildings to demonstrate the effectiveness of their joinery

7-10: *A two-story bamboo structure by Marcelo Villegas.*

system. And interest in bamboo construction dramatically increased because of the positive performance of bamboo structures during the major earthquake in Colombia in January 1999. (When I visited the Guadua bamboo region of Colombia several months before that earthquake, I was shocked to see how little was being built of bamboo, except in the rural areas. Now there are perhaps 30 or 40 professionals designing with bamboo in the region.)

7-11: *This detail demonstrates the bolted connection with concrete infill that has been instrumental in creating the latest innovations in bamboo architecture in Colombia.*

The work of the next generation has started to demonstrate new esthetic and engineering approaches. Until now each designer has had to find willing patrons with whom to make mistakes. But in the 1999 earthquake, 75 percent of the masonry buildings collapsed, while virtually all of the bamboo structures survived. The wealthy now consider bamboo an acceptable building material and are allowing some interesting experimentation to go on. (And where the wealthy lead, the masses are soon to follow.)

Architects and designers wanting to work with bamboo are actively working with structural engineers to refine the formulas and practice of structural design. Although one of Colombia's leading designers, Raphael Rojas, readily admits that some of his early designs have required supplemental work, his current work is using bamboo to its potential. Another architect, Saúl Vera (who helped Rojas with the building of his early works) designs/builds with his own crews, spreading the gospel of bamboo. And architect Hector Silva has built some very elegant and simple social housing, and a rural school, whose designs draw on traditional *bahareque* wall-building techniques and the trusses of Vélez.

And that's not all. A committee led by Dr. Jules Janssen, under the mantle of INBAR (the International Network of Bamboo and Rattan), has put together a manual for the standardized testing needed to compare different bamboo species, to compare one person's testing with another's, and to apply those numbers to engineering formulas. In Colombia this challenge has been met by the work of several universities and the Association of Seismic Engineers, who have put *Guadua Angustifolia* through the whole range of tests. The development of the manual may make it possible for Colombia to export timber bamboo to the US, since the standard adopted here requires the same battery of tests.

And for the first time in at least 500 years, there is code approval in Colombia for *bahareque*, the wall system that uses wood or bamboo studs and flattened bamboo *esterilla* for lathing before plastering. Manuals hot off the press explain the best ways to build bamboo structures and how to avoid traditional detailing mistakes. The manuals are highly illustrated (so literacy is not required) and can work well on other continents.

German bridge designer Jörg Stamm is pushing hardest at the limits of the material. Living in the bamboo region of Colombia, Stamm has built bamboo bridges with clear spans as long as 165 feet (50 meters). He was initially aiming for spans half as great, but realized the possibilities when calculations showed they could work. Stamm has been very generous with his knowledge, writing a book that shares his five years of intense research on how to build bamboo bridges. Stamm feels strongly that the use of the bamboo can save some of the large rainforest trees. His only limitation now seems to be the state of the Colombian economy as he searches for new projects.

Leading practitioners also learn from each other's work. Several years ago Vélez told me that he had no aspirations to ever create curved bamboo structures. He thought he had a lifetime's worth of rectilinear ideas still to explore. Yet he couldn't help but be inspired by the sensual curves in the work of Raphael Rojas. And, sure enough, when Vélez was asked to design a bamboo cathedral to replace the collapsed brick and concrete one, almost every piece was curved. The transparent façade allows light to filter into the interior more subtly than the best of Europe's cathedrals. In this example, as in so many others, the new Colombian bamboo architecture demonstrates the power of working with a simple material, together with the collective knowledge of the local people that leads to the best use of it, resulting in sustainable structures of strength and grandeur.

7-12: *This bamboo bridge by Jörg Stamm has a much greater span than previously thought possible. Stamm's rigorous structural work stretches the boundaries of bamboo architecture.*

7-13: *A dramatic bamboo pavilion by Simón Vélez.*

Case Study:
Natural Building in Poland

Flora Gathorne-Hardy

Paulina Wojciechowska was born in Poland and lived there until she was ten, when her family relocated to Afghanistan. There she first encountered the beauty of vernacular building. Twenty years later at a seminar in Warsaw, Wojciechowska (by then a trained architect with earth-building experience in the UK, US, and Mexico and the founder of the nonprofit "Earth, Hands, and Houses") decided to design and build a load-bearing straw bale house that could act as a demonstration project in her home country. Many traditional buildings in Poland are constructed from earth and other local materials, and there was a small but growing interest in applying ecological design techniques to new buildings. What better way could there be to inspire others?

The Location

Wojciechowska was drawn to the northeast of Poland, to a family farm in Prselomka that doubled as a gallery and educational center. The location was ideal. Przelomka, in the Suwalszczyzna region of Poland, boasts some of the most beautiful and unspoiled countryside that Poland can offer. Indeed it is known as the "green lungs" of the country. Unpolluted lakes teem with fish, woods are rich in fauna and flora, and low-intensity farming supports a richness of invertebrate and bird life. Characterized by warm summers and cold winters, it was the perfect place to focus attention on construction techniques that use materials that are right at hand. And the new buildings would attract people to the gallery, thus helping to support the local economy in an area that had seem many changes under the rule of different political regimes. The rural economy consisted of small family farms, many of which provided little financial income but retained great cultural wealth in traditional craft, farming techniques, and animal husbandry.

Design and Construction

The house was located in a sloping meadow alongside a small lake and close to existing farm buildings. It was designed to accommodate a family and has a kitchen, sitting room, bathroom, hallway, and two upstairs bedrooms reached by a staircase. The building is rented to tourists who are interested in exploring the area from the base of a truly ecological home.

The house was built over three summers, using volunteer labor. Local people were employed to demonstrate as many of the traditional skills as possible: roof framing using methods particular to the area, wheat-straw thatching, and the traditional house stove. Construction materials came from no more than half a mile (one kilometer) from the site. The foundations were made of field stones; straw bales came from the neighboring land; wood came from storm-felled trees; and clay came from a hole directly next to the house. External seating was created from stones gathered using horse and cart, and a shaded canopy was built from coppiced hazel harvested by local children.

During the construction process, media attention grew to the point that backpacking volunteers would appear each day, willing to help. Many hands sculpted internal walls, created smooth finishes on the heated seat, or invented exciting forms that curl along the hazel partitions. Children left their mark, too, helping to mold seats to their own forms and enjoying the sensations of sifting and applying fine finishes. Even traveling musicians took inspiration from these artistic shapes, playing tunes to celebrate the new house and stamp down the earthen floors.

7-14: *This natural house in an unspoiled part of Poland is an example of how a region can build upon its traditional heritage for the benefit of local residents, without destroying the surrounding ecosystem.*

Taking Stock

The building of the house has performed many different functions, some unexpected. On the most basic level, it has provided a new source of income for a rural family who is able to advertise it as an ecologically designed building. This is significant because, for this and other rural regions in Poland, ecotourism offers a new and crucially important source of revenue. Poor soils, small farms, and falling prices for meat and foodstuffs have all combined to make it increasingly difficult for anyone to make a living from the land. A steady outflow of people who do not wish to leave are forced to become economic migrants, seeking work in the cities or abroad. Yet, at the same time, their simple way of farming has safeguarded the ecology of the area (storks, for example, breed in great numbers and fields to remain rich in colorful flora). Ecotourism builds on these strengths, showing how farming and the environment do indeed need to travel hand in hand. Rather than being seen as backward, such low-intensity methods of farming and these similarly low-intensity forms of building are at last being seen as models for the future.

Within the local community, the house has continued to attract attention as a locus for performances and social events. And there is a continuing program of building workshops.

So far, participants have constructed a dining shelter and a stove for baking traditional bread. Older people take pride in their work on the building and share their knowledge with younger people. And those with traditional skills feel confident that their work is of value.

Looking wider still, what is exciting is the way the project has spurred action across the country. Hundreds of people have come to visit the site, many of whom attended workshops or simply found themselves offering a helping hand. These experiences have helped generate ideas, and now the list of new projects inspired by the house grows longer every week. A proposal is in the works to build a straw bale stable designed to maximize the welfare of the horses. Two former workshop attendees are building a straw-clay block kindergarten outside Warsaw (see "Straw-Clay Blocks"). And ecotourism projects are getting underway in the southern regions of Poland. That it is possible to build a relatively low-cost house from simple materials has offered some people an escape from homelessness and provided them with the security of a place of their own.

Looking Forward

Media interest in the house and the techniques it demonstrates continues to grow even now, more than a year after its completion. Television, the national press, and radio journalists are paying more and more attention to the ideas and practical benefits of ecological design. They use the house as a case study that people from all walks of life can understand. The sculptural beauty of the building, the comfort of the heated bench, and the colors of the earth walls all combat any negative stereotyping of ecological design as poor design or backwardness. Instead people see a home of simple beauty and tremendous spirit, lodged at ease in the landscape and offering promise of a more sustainable and humane future for us all.

PAULINA WOJCIECHOWSKA

7-15: *Children mixing earth plaster for use on a straw bale house.*

Case Study:
Low-Cost Housing Projects Using Earth, Sand, and Bamboo

Gernot Minke

Introduction

Earth, sand, and bamboo are environmentally friendly building materials that are locally available in many parts of the world. They do not require industrialized processing (which pollutes the environment) nor do they need extensive energy, capital, or high levels of skill to use. The Building Research Institute (BRI) of the University of Kassel, Germany has done extensive research into the properties of earth, sand, and bamboo and has implemented low-cost housing projects using innovative building methods using these materials in Guatemala, Ecuador, Chile, Bolivia, and India. These projects show that it is possible to build earthquake-resistant walls from rammed earth, sand-filled cotton bags, or from bamboo — without using reinforced concrete elements. Furthermore it is possible to built hurricane-resistant vaults and domes from adobes covered with earth and vegetation, which can be built by owners with little expert advice.

Lessons from Earthquakes

I visited Guatemala in 1977 after an earthquake there killed more than 20,000 rural dwellers. Most were killed within minutes of the onset of the quake as the brick or adobe walls of even single-story houses failed, typically falling outward. Once the wall support was gone, the heavy tile roofs fell, killing the people who slept underneath them.

International aid services built new houses from steel, plywood, asbestos cement, or concrete slabs. None considered using the traditional building material — earth. They all thought that it was impossible to built earthquake-resistant houses of earth. According to our research, however, it was not the building material that caused the collapse but improper building techniques.

7-16: *A model of an adobe house after testing on a shake table, illustrating typical design mistakes.*

A "shake table" test of a house with adobe walls shows some typical design mistakes: diagonal cracks arising at the corners of a window, the wrong lintel length, and the absence of a ring beam, to name a few. (There are about ten general mistakes that might lead to collapse by seismic shocks.) When I studied earthquake damage in Mendoza, Argentina I learned that it is no longer permissible to build earthen houses. Despite this prohibition, 80 percent of the rural inhabitants of the region still rely on earth, as they cannot afford other, more expensive materials. Yet in this same area, I found old houses with thick, rammed earth walls that had withstood all the earthquakes of the last two centuries. Knowing that earth was still viable and realizing that most rural people could not afford the methods prescribed by the aid agencies, I decided to develop earthquake-resistant houses using earth as the building material.

Bamboo-Reinforced Rammed Earth Wall System

BRI developed and successfully implemented an earthquake-resistant prototype house in Guatemala in 1978. We erected a rammed-earth wall over a bamboo-reinforced mono-lithic stone foundation, using a special T-shaped steel slip form. The 31-inch-wide (80-centimeter-wide) wall sections are rammed continuously until the final wall height is reached, so that no horizontal shrinkage cracks can occur. They are separated by a ¾-inch (2-centimeter) gap (which is later filled by earth or moss), to provide for independent movement during seismic activities.

Four vertical bamboo rods act as reinforcement for the wall sections, and together with the gap give sufficient stability against horizontal loads created by seismic shocks. The bamboo is interconnected with the reinforcement of the foundation, as well as with the ring beam that connects the tops of all the wall sections. The wall surface is protected against rain by a painted-on lime solution. The roof rests on six wooden posts that stand 20 inches (50 centimeters) inside the rammed earth wall. This is an essential solution for earthquake resistance, as roof and walls have different mass, acceleration, and frequency during earthquakes and should be able to move independently.

BRI built another reinforced rammed-wall system in 2001 at Alhué, Chile. In this case the walls were stabilized by their L - or U-shape and had additional 2-inch-thick (5-centimeter-thick) vertical inner elements of a local reed similar to bamboo. Again the roof rests on posts independent of the wall system.

7-17: A low-cost bamboo-reinforced earth house project in Guatemala.

Elemented Rammed-Earth Wall System

In Pujili, Ecuador, together with FUNHABIT, Quito the author built a low-cost housing prototype, using the local clayey soil mixed with pumice. The stabilization effect of the walls was created by their L-shaped elements that are interconnected at the top by a ring beam. In this case the posts of the roof rest outside of the walls to allow independent movement during earthquakes. The roof was built of eucalyptus trunks and local reed covered by a mix of earth, pumice, animal dung, sisal fibers, and waste car oil.

Earth-Filled Hoses for Walls

In Guatemala, a second low-cost housing prototype was developed by BRI in 1978. Cotton fabric hoses up to 9 feet (2.8 meters) long were filled with earth and pumice, dipped into a solution of lime to prevent rotting, and then stacked. The walls were stabilized by wooden posts placed 7 feet (2.25 meters) apart and by thin bamboo rods every 18 inches (45 centimeters). At the top these vertical elements were fixed to a horizontal ring beam. This system gives sufficient flexibility to withstand earthquake movements. The material costs inclusive of doors, windows, etc. were US$610 for a 59-square-foot (55-square-meter) house ($10 per square foot; $11 per square meter). Additional labor costs were only about 20 percent of the materials cost.

Bamboo Prototype Building

During a training course for architects and engineers held by the author at Babahoyo, Ecuador in 1989, we demonstrated methods of constructing roofs and walls from bamboo. To reduce deterioration of the bamboo from insects and climate, the bamboo pieces were cut in half, dipped into waste car oil for impregnation and added to prefabricated roof elements. [Editors note: using waste car oil is potentially toxic to inhabitants, especially when used for interior elements.] Bamboo rods were also used in this project to build rigid walls.

Adobe Vaults and Domes

The author developed a bamboo-reinforced earthen vault system in 2001. Special adobes were attached with mud mortar to a catenary arch form built of three layers of split bamboo, which stays in place. Although in a heavy earthquake the mortar joints may crack and the vault might move due to the elastic behavior of the bamboo arches, the kinetic energy of the seismic shocks is absorbed by the deformation of the flexible form-work, which, despite deformation, can continue to support the adobe vault.

In La Paz, Bolivia in 2000 the author built an earthquake-resistant adobe dome using a rotational guide developed by BRI. The dome has an optimized cross-section, calculated by a computer program that guarantees that no tensile forces (which could lead to cracking and dome failure) occur within the dome. The foundation, a circular reinforced concrete ring beam, takes the shear forces of the dome. This dome has a free span of 30 feet (9 meters) and a

GERNOT MINKE

7-18: *Bamboo prototype building in Babahoyo, Ecuador.*

height of 20 feet (6 meters). The adobes have a special curved surface, with rounded corners in order to give a good sound quality, and the joints are deepened to give some sound absorption effect. Several buildings have been built in Germany and India using this technique. The domes are covered by a 6-inch (15-centimeter) layer of earth on which wild grasses are grown. This "green roof" gives a high cooling effect in summer and a good thermal insulation effect in winter. The earth layer also gives a structural stabilization effect against asymmetric loads.

GERNOT MINKE

7-19: *The interior of an adobe house in La Paz, Bolivia. Note the articulation of adobes on the inner surface for better acoustic performance.*

SMALL-DIAMETER WOOD:
AN UNDERUSED BUILDING MATERIAL

Owen Geiger

Many of those in need of housing have access to small-diameter trees in nearby forests. These trees can be used to produce materials that are ideal for building affordable homes. If used in conjunction with energy-efficient straw bale construction and other natural materials, small-diameter wood can be used to create a better home than most building systems, at lower cost.

US forests have been poorly managed and are now choked with small-diameter trees. Thinning these trees to reduce the risk of forest fires, which is currently at a record high, is a Forest Service priority. With an inexpensive firewood permit, anyone can obtain small-diameter wood for building a home. (Even though this wood could be used for firewood, it is more valuable as a building material.)

There are several advantages to using small-diameter wood for building:

- Small-diameter wood can provide all of the lumber for a house including studs, joists, plates, trusses, window and door frames, trim, and other components. Wood frame construction is the preferred building system in the US because of its speed and ease of construction, but unfortunately most dimensional lumber is not sustainably harvested. Using small-diameter wood encompasses the advantages of wood frame construction but uses wood that improves the health of the forest and reduces forest fires.
- The use of locally available wood reduces construction costs and avoids supporting environmentally irresponsible lumber companies.

- Wood in the round is much stronger than standard dimension lumber and requires less processing. Thus smaller diameter logs can be used, with fewer parts. For example only one-half as many trusses may be required, because pole trusses can be set every 48 inches (122 centimeters) instead of every 24 inches (61 centimeters).
- A US$40 chainsaw guide can be used to mill purlins, joists, studs, plates, and other components. (The Beam Machine is one example of a low-cost chainsaw attachment that can mill straight edges on poles.)
- The fire resistance of poles is much higher than stick-framed trusses or engineered trusses (TJIs). Wood poles have a two-hour commercial fire rating, in contrast to the other two options, which have a one-hour fire rating. And in the event of a fire, there is no toxic off-gasing — the leading killer in home fires.
- Timber frame/pole construction is more esthetically pleasing than wood frame construction covered with plasterboard. The beauty of the wood is left exposed, honoring the tree from which it came.
- Very few tools are required to build simple pole trusses. If they are built in uniform sizes, workers can be trained to build them quickly.

KELLY LERNER

7-20: *Small-diameter wood poles create the structure for a straw bale house in Argentina.*

The use of small-diameter wood creates local jobs and places less reliance on highly processed materials that must be shipped long distances. Jobs are created in four categories:

1. Logging: Workers are needed to cut, mill, and deliver poles.
2. Truss manufacturing: Workers are needed to build roof trusses. This could be a cooperative effort or an entrepreneurial cottage industry. Either way the quality will be higher and more consistent if specially trained workers build the trusses.
3. Milling: Workers are needed to mill logs into purlins, studs, plates, or joists. The simplest method uses a chainsaw and a guide. Mass production methods with commercial-sized equipment are even faster and more efficient.
4. Construction: Workers are needed to erect trusses, build walls, etc.

With all the advantages of small-diameter wood, we should take a closer look at how to use this resource that is so often near at hand.

CHUCK MARSH

7-21: *This example of round poles used to create an octagonal roof structure demonstrates the beauty of this locally available material. Using round poles in some situations, however, may demand a high level of craftsmanship for good structural connections and esthetics.*

Builders Without Borders Training Programs:
An Educational Strategy for Sustainable Construction

Edited by Owen Geiger, Builders Without Borders

Builders Without Borders (BWB) envisions humankind working together in a "global village" to alleviate substandard housing and poverty. BWB continues to share lessons learned with the network of builders and development organizations that are working to alleviate substandard housing by providing workshops, training programs, and educational materials. Ultimately the solution to the world's housing crisis lies in training and education.

Builders Without Borders has four distinct training programs in various stages of development:

1. Facilitator Training;
2. Workshops;
3. In-house Home Construction Training in the US (Phase I); and
4. In-country Home Construction Training (Phase II).

BWB Facilitator Training: Teaching the Teachers

BWB quickly realized that we would not help solve the housing crisis by simply becoming another provider of housing. The Facilitator Training course focuses on teaching teachers of natural building how to work within communities in a culturally sensitive and effective manner. The goal is not so much about finding answers as about asking the right questions, so that best practices can emerge from collective knowledge and then be effectively developed and replicated by local builders. By focusing on the goal of "teaching teachers," we seek to create a process-oriented approach to developing local housing solutions that will come from within each individual community. That way the appropriate designs stand a better chance of being replicated without outside support.

The Facilitator Training course includes a lively discussion on the many complex issues surrounding "development" work. Within the context of a hands-on building project, the course goes beyond technique to broaden perspectives about working within other cultures and to identify those areas where teachers can be of value. Trainers with international development experience share lessons learned from their work. Attendees are encouraged to share what has worked for them in a variety of circumstances, as well as what has failed. These discussions lead to an honest and deep exploration of both concerns and knowledge base by a wide spectrum of participants.

The training covers climate and culture-appropriate design; hands-on experience with a variety of natural building techniques; and skills to enhance working in the field, including logistics, communication, cultural sensitivity, and other topics. Those who successfully complete the training are encouraged to begin developing their own projects of interest. BWB plans to refine the Facilitator Training course so that it can eventually be replicated in different regions around the world, several times annually.

Builders Without Borders Workshops

An overarching goal of BWB is to empower people rather than just telling them how things should be done — something that might prevent many housing projects from failing. The Chinese proverb, "If I hear it, I forget it. If I see it, I remember it. If I do it, I know it." is quite appropriate in our work. An emphasis on training, including job training for local communities, rather than just building houses is one of the keys to successful projects. Experiential learning is best. It gets across key points and helps people recognize why they're doing what they're doing. BWB workshops encourage shared learning, self-discovery, hands-on activities, open dialog, problem solving, and the building of multicultural relationships.

BWB workshops include both theory and practice and emphasize collaborative learning through hands-on experiences. Participants learn how to save time, labor, and materials using the latest methods. A typical lesson includes discussions and visual training aids such as slides and handouts. Interactive exercises and demonstrations at the building site encourages participants to immediately begin applying what they have learned. Participants gain experience in all phases of construction by working closely with master builders on the construction of an actual home or other structure.

Instructors serve primarily as facilitators to lead the group and help participants learn through an interactive, participatory process that encourages creative thinking, enthusiasm, sharing, and reciprocal learning. We use group discussions instead of a static lecture format and encourage participants to ask questions, talk about their experiences, and make suggestions. This free exchange of ideas adds to the continual development and improvement of the building process. Learning and implementing local skills and knowledge is an important part of the process because we realize that we have as much to learn (if not more) as we do to teach.

BWB In-house Home Construction Training in the US (Phase I)

The Home Construction Training (HCT) program provides home construction training for architects, builders, community planners, and instructors using low-cost, natural building methods. Phase I is a two-month program that takes place in New Mexico, where BWB has ready access to trainers, staff, and materials, as well as the extensive US network of natural building professionals. Participants acquire the necessary skills, knowledge, and experience to contribute directly to the building of homes in their region. (We believe that this is the most effective approach to solving the housing crisis, because it is empowering and fosters independence from outside influences.) Written and video documentation, with portions in English and in local languages, provides a lasting reference of these building technologies. Information can then be taken home, distributed, and implemented throughout the affected region. And it can be customized to meet the needs of specific regions and countries.

Phase I focuses on the collaborative research and prototype design work that is required to obtain the most cost-effective home designs. BWB works closely with architects, planners, and builders to develop locally appropriate house designs and building systems. Design charrettes, demonstrations, group exercises, and brainstorming sessions help identify the most promising solutions. Participants then work hands-on with the materials, develop techniques, and refine the prototype designs along with experienced US professionals to create an effective process and program.

App-1: *Members of Builders Without Borders in front of the training structure built during its first Project Facilitator training in Kingston, New Mexico.*

Primary Teaching Methods

Primary teaching methods include visual presentations and collaborative learning through hands-on experience. A typical lesson includes a brief period of classroom instruction (with a talk, slides, handouts, discussions, etc.), followed by demonstrations at the building site where participants immediately begin applying what they have learned. Participants thus gain experience in all phases of construction by working closely with master builders on the construction of a prototype structure.

Instructors serve primarily as facilitators to lead the group and help participants learn through an interactive, participatory process that encourages creative thinking, enthusiasm, sharing, and reciprocal learning. Group discussions encourage participants to ask questions, talk about their experiences, and make suggestions.

Leading building professionals are routinely incorporated into the training as guest instructors to provide in-depth knowledge in specialized areas of expertise. Participants also gain knowledge from tours, collaborative projects, electronic resources, library resources and networking with others. The tours enable participants to see firsthand how low-cost building materials are being employed to build high quality homes in the US.

Training Aids

Quality training aids can significantly increase the level of learning and are particularly important for overcoming language barriers. The highly graphic training aids used in Phase I include slides, videos, posters, handouts, blackboard presentations, and line drawings. Each participant receives a 3-ring notebook, with instructional materials translated into the appropriate language (translation into additional languages can be arranged). A bibliography, a resource list of materials and supplies, and video documentation is included as part of the course.

Video documentation is one of the most valuable components of this program, because the visual format communicates across languages and literacy boundaries. And the video can be extremely useful for in-country training, especially when native speakers who are HCT participants are filmed explaining key concepts. After the completion of the training, the video is edited and sent to the client. The video documentation is available on tape and DVD, with multiple copies and formats as agreed upon to meet the participants' needs.

Topics Covered in Phase I

Topics covered in Phase I include:

- Natural building
- Passive solar design

- Seismic design
- Straw bale construction
- Adobe and straw-clay construction
- Alternative foundations
- Roofs
- Earthen floors and plasters
- Solar ovens and water heaters
- Water catchment
- Waste disposal
- Construction management
- Culturally integrated design
- Collaborative design processes
- Community development
- Other housing issues.

Order of Activities

The two months required for the completion of Phase I are organized into four seg-
ments, including:

- An orientation and technology tour (two weeks). Participants are introduced
  to the program, the facilities, and the fundamentals of sustainable development and natural building. A technology tour explores projects in New
  Mexico and southern Colorado, which are hotbeds of innovation in alternative building. Day trips to these sites enable participants to experience a rich
  variety of exemplary projects including straw bale, adobe, rammed-earth
  tires, pumicecrete, straw-clay, cob, plasters, papercrete, earthbags, intentional
  communities, etc.
- A hands-on technical training overview (one week). Participants share the
  design parameters that are present in their country and are exposed to natural building methods that would be appropriate for their specific area.
- A design brainstorm (one week). Site-specific building designs are developed
  with experienced engineers, planners, architects, and builders. The designs
  reflect the needs and conditions of the various climate regions within the
  country. The inclusion of building professionals and community leaders
  (both male and female) from various regions within the country is recommended for this process.

- A construction project /in-depth technical training (four weeks). Participants work alongside instructors to build an entire structure using materials that are locally available in their country. This is the primary activity of the training program and requires extensive effort on behalf of the participants.

Interpreters

A minimum of two interpreters is necessary to ensure clear communication at all times. For instance, each primary instructor requires an interpreter during construction on the job site. If possible, the provision of additional interpreters is encouraged to improve the training process.

App-2: A straw bale and round pole hogan, built with volunteers from BWB, the Civil Air Patrol, and Native American youth. This project demonstrated the importance of adequate communication between culturally different groups, as well as the challenges of building an unusually shaped building (in this case, a hexagon) with unskilled labor.

Selection of Participants

Ideal candidates should have extensive experience in residential construction and/or architectural design and be enthusiastic about natural building. They should possess communication and leadership skills that enable them to transfer information to others. Participants with a broad background in all aspects of construction are preferred over those with skills in only one area. Candidates must be genuinely concerned with seeking affordable housing solutions for low-income families and be willing to work within their communities. We favor participants, both men and women, from diverse regions, and multicultural backgrounds so that the building techniques covered in the program become more widely adopted. Women in particular are often highly motivated and dedicated to housing issues, and we encourage their inclusion as participants.

In summary participants of this training will receive:

- Customized home designs for various regions in their country, with input from professionals in the field of natural building
- Hands-on building experience
- Video documentation in their own language for use as a training tool
- Instructional materials
- Personal relationships with professionals for future consultation
- New personal and leadership skills that foster independence.

BWB In-Country Home Construction Training (Phase II)

Phase II of the Home Construction Training program (HCT) is a longer, more comprehensive housing plan that builds on the training from Phase I. Phase II focuses on training builders in their home country and on building houses with the support of Builders Without Borders advisors. Phase II uses the designs, construction methodologies, and graduates from Phase I for the most effective results. Phase II can proceed immediately after Phase I, enabling participants to begin applying their new skills as soon as possible.

The ambitious program is intended to build a large number of homes while training and empowering local communities to take over this function for themselves. Local builders carry on the program as they complete the training. The program expands into other communities over time, developing local skills and creating micro-enterprises such as pallet truss and adobe block manufacturing. It also allows for partnering with other organizations to address and incorporate solutions to other human needs such as health and education.

App-3: *Kelly Lerner and Chinese builders in front of one of the projects built in China under the auspices of the Adventist Development and Relief Agency.*

Home designs are chosen by the future homeowners, with allowances made for individual adaptation and creativity (thus avoiding sterile "cookie-cutter" communities). The designs use local materials as much as possible and encompass a range of building methods such as straw bale, adobe, and earthbag construction, etc., as deemed appropriate by each community. Construction sites are user-friendly because of the low-tech natural building approach, which encourages unskilled workers, curiosity seekers, women, children, and other community members to participate. Ultimately their involvement speeds the adoption of these building methodologies throughout the area.

A community planning team works with Phase I graduates and local community elders to develop suitable town plans. These may include building on old foundations or designing a new community infrastructure that addresses water, sanitation, energy, gathering places and other public structures. The first building to be built (a community center) serves as the BWB headquarters for the duration of the training cycle. Later it can serve as a local community clinic, school, library, town hall, or other structure. Working together on this structure brings people together for a common cause, and helps build relationships in the process.

Training Aids

The videos created in Phase I are the primary training aids used in Phase II. The videos show each step of construction for each house design and can be updated as needed to reflect any changes in design. Other visual and written documentation developed during Phase I is used, as well. As the designs are adapted to local conditions and tested, the principal trainers develop written training materials into a "Training Guide." Not exactly a how-to manual, the Training Guide is continuously updated as new ideas and improvements emerge. It contains lessons that cover each step of construction, from laying out the foundation to building the roof. Line drawings and pictures help bridge language barriers. The Training Guide can be translated into any language and made available on CD-ROM.

App-4: *The beauty of building in community: an adobe brick chain in Thailand.*

Although final house designs must be developed through collaboration with local building professionals, a few potential solutions are described here in order to point out some of the benefits of natural building:

- Straw bale construction with passive solar design offers excellent insulation so that houses stay warm in the winter and cool in the summer. Straw bale houses are affordable, fire resistant, fast and easy to construct, and resistant to insects and vermin once properly plastered. Straw bale houses can be engineered for earthquake resistance. They also require minimal use of lumber and other costly materials.

- Straw-clay construction combines the advantages of adobe construction (such as high mass) with some of the advantages of straw bale construction (higher insulation, lighter weight walls).

- Vaults and domes (for desert regions), created from earth, straw-clay, earthbags, or stone can be enhanced with engineered foundations, additional insulation, and other improvements. Vaults and domes are very resource efficient because they do not require wood, steel, or other structural roofing components in their construction.

Construction Team

A team approach is used in the construction of homes. The focus is on training the apprentices so that they acquire the necessary skills to contract houses in the future. Family members contribute to their future home, and volunteers gain valuable skills, as well.

Homeowners

Homeowners assist in the building of their own homes by working directly with the work crews. The benefits of self-build are widely acknowledged, and this approach is an integral part of the program. For example homeowners develop a sense of ownership and pride, learning how to maintain their homes and make future repairs. Homeowner participation lowers costs and improves the quality of the houses. And the construction process is shorter with the help of family members and community volunteers.

Volunteers

Volunteers from the community are encouraged to participate. This additional help speeds up the construction process, lowers costs and empowers the local people directly. Volunteers gain valuable experience that can enable them to build houses for themselves without further training. Since building materials are ideally mostly locally available, many of these volunteers will be able to begin building on their own after participating in the building process. Evening presentations in the community building also help volunteers learn the necessary skills. Lunches are provided for volunteers to encourage their participation.

Graduates

After completing the training apprentices will be qualified to build houses on their own from start to finish. These graduates will be able to build new houses with the aid of a crew, homeowners and volunteers. Graduates acquire and supervise their own crews, and bid for new projects on a competitive basis to meet the needs of the community.

Facilities and Regional Information

Training sites are located in various areas of the host country. Sites are carefully chosen so that advisors, building materials, tools, office equipment, supplies and other resources can be readily transported from other areas. New sites are chosen each year to spread the program to other communities, and as more funding becomes available, the program will be expanded into more sites.

Interpreters

A minimum of two interpreters per region is necessary to ensure clear communication at all times. For instance, each principal instructor requires an interpreter during construction on the job site. The cost of two interpreters per region is included in the budget. Additional interpreters would be beneficial to the training process.

Management

Experienced managers of Builders Without Borders will provide oversight of the entire program from curriculum development to daily training sessions. Builders Without Borders will also draw from other experienced US designers, builders, and community planners, as well as its worldwide network of natural building professionals to bring an international presence to the process.

Conclusion

We believe that the most effective way to provide affordable and comfortable housing is to make use of low-cost, locally available materials because other alternatives are prohibitively expensive, impractical, or simply unavailable. We are also convinced that training a local workforce is the most productive and sustainable approach to providing affordable housing. The most effective way to train a skilled workforce, in our opinion, is with the training program described here. After the completion of this training, home designs and skills will be in local control. Participants will then be able to guide the local programs, with less reliance on foreign assistance and expertise.

Notes

Chapter 1: Shelter and Sustainable Development

1. See Shelter for All: Global Strategy for Shelter to the Year 2000 (United Nations Centre for Human Settlements, 1990) [UN-Habitat Publication reference HS/195/90F and HS/197/90F].

2. See United Nations Centre for Human Settlements (UNCHS) website at: www.unchs.org [cited October 2000].

3. A United Nations Development Programme (UNDP) statistic, quoted in Nicholas Low, Brendan Gleeson, Ingemar Elander, and Rolf Likskog, Consuming Cities: The Urban Environment in the Global Economy after the Rio Declaration (Routledge, 2000), p. 8.

4. See UNCHS website [cited October 2000].

5. See Molly O'Meara, Reinventing Cities for People and the Planet (Worldwatch Institute, 1999), p. 15.

6. See Low, Gleeson, Elander, and Lidskog, 2000, p. 11.

7. See "Istanbul Declaration on Human Settlements" (UNCHS Strategy Report, January 1996), UNCHS website [cited June 2000].

8. See Low, Gleeson, Elander, and Lidskog, 2000, p. 11.

9. See Annex I, Paragraph 61 of the "Habitat Agenda," UNCHS website [cited June 2000].

10. See UNCHS website [cited June 2000].

11. See Athena Swentzell Steen, David A. Bainbridge, and Bill Steen, The Straw Bale House (Chelsea Green Publishers, 1994), p. xii.

12. See Lloyd Kahn, Shelter (Shelter Publications, 1973), p. 3.

13. See James Steele, Sustainable Architecture: Principles, Paradigms, and Case Studies (McGraw Hill, 1997), p.58.

14. See Low, Gleeson, Elander, and Lidskog, 2000, p. 7.

15. See Wolfgang Sachs, "The Need for the Home Perspective," The Post Development Reader, Majid Rahnema and Vitoria Bawtee, eds. (Zed Books, 1997), p. 298.

16. See Wolfgang Sachs, ed., The Development Dictionary: A Guide to Knowledge as Power (Zed Books, 1997), p. 2.

17. See World Commission on Environment and Development, Our Common Future (Oxford University Press, 1987), p. 43.

18. See Gustavo Esteva, "Development" (1997), The Development Dictionary: A Guide to Knowledge as Power (Zed Books, 1997), p. 10.

19. See Nabeel Hamdi, Housing without Houses: Participation, Flexibility, and Enablement (Van Nostrand Reinhold, 1991) .

20. See Laura Macdonald, Supporting Civil Society: The Political Role of Non-Governmental Organizations in Central America (MacMillan, 1997), p. 25.

21. See Alan Gilbert and Josef Gugler, Cities, Poverty, and Development: Urbanization in the Third World (Oxford University Press, 1992), p. 115.

22. See Simone Fass, "Housing the Ultra-Poor: Theory and Practice in Haiti," JAPA Journal of the American Planning Association, Vol. 53 No. 2 (American Planning Association, 1987), p. 198.

23. See Gilbert and Gugler, 1992, p. 119.

24. See Low, Gleeson, Elander, and Lidskog, 2000, p. 4.

25. See John Friedmann, Empowerment: The Politics of Alternative Development (Blackwell Publishers, 1992), p. 23.

26. See Hamish S. Murison and John P. Lea, eds., Housing in Third World Countries: Perspectives on Policy and Practice (St. Martin's Press, 1979).

27. See Michael L. Leaf, Urban Housing in Third World Market Economies (Asian Urban Research Network, UNCHS, the Canadian International Development Agency (CIDA), and the University of British Columbia, 1993).

28 See Murison and Lea, 1979, pp. 126-128.

29. See UNCHS, Shelter for All, 1990.

30. See Friedmann, 1992, p. 107.

31. See Caroline Moser and Linda Peake, Women, Human Settlements, and Housing (Tavistock Publications, 1987), p. 13.

32. See John F. C. Turner, Housing By People: Towards Autonomy in Building Environments (Pantheon Books, 1976), pp. 61-64.

33. See Gilbert and Gugler, 1992.

34. See UNCHS, Global Report on Human Settlements (Oxford University Press, 1987), p. 76.

35. See UNCHS, Shelter for All, 1990.

36. See UNCHS, Shelter for All, 1990.

37. See UNCHS, Global Report on Human Settlements, 1987, p. 175.

38. See Steele, 1997, pp. 13, 16.

39. See Ismail Seregeldin, The Architecture of Empowerment: People, Shelter, and Livable Cities (The Academy Group, 1997), p. 29.

Profile: Sri Laurie Baker, Architect

1. Laurie Baker, "Baker on Laurie Baker Architecture," u.p., quoted in "Baker of India," Gerard Da Cunha, Architecture and Urbanism (A+U) 363 (A+U Publishing, December 2000).

2. See G. Bahtia, Laurie Baker Life, Works, and Writings (Penguin Books, 1991).

3. See Note 2.

4. See Note 2.

5. See Hassan Fathy, Architecture for the Poor (University of Chicago Press, 1973); and Natural Energy and Vernacular Architecture (University of Chicago Press, 1986).

6. See Laurie Baker, Low-Cost Buildings for All, Hindustan Times, New Delhi, 17 January, 1974.

7. See Note 1.

Chapter 2: Speaking the Vernacular: Mud versus Money in Africa, Asia, and the US Southwest

1. See Ivan Illich, Toward a History of Needs (Pantheon Books, 1978); William Leiss, The Limits to Satisfaction (University of Toronto Press, 1976); Marshall Sahlins, Stone Age Economics (Tavistock Publications, 1974); and Karl Polanyi, The Great Transformation (Ocatgon Books, 1975).

2. See Marshall Sahlins, p. 13.

3. On "development," see Ivan Illich, Shadow Work (University of Cape Town, 1980), pp. 15-19. The biologist Paul Erlich calls the industrialized world "overdeveloped" and

needing "de-development"; quoted in John H. Bodley, Anthropology and Contemporary Human Problems (Mayfield Publishing, 2001), p. 226.

4. See Lewis Mumford, The Myth of the Machine, Vol.2 (MJF Books, 1997), pp. 400-404.

5. See John. H. Bodley, Victims of Progress (Mayfield Publishing, 1999), p.130.

6. Another important kind of non-standard construction would be owner-building. See Ken Kern, The Owner Builder and the Code (Owner-Builder Publications, 1976).

7. See George Dalton, "Theoretical Issues in Economic Anthropology," Studies in Economic Anthropology (American Anthropological Association, 1971).

8. For eloquence on the subject, see Shelly Kellman, "The Yanomamis: Their Battle for Survival," Journal of International Affairs, 36:1 (Spring/Summer, 1982), pp. 15-42. Toward analysis, see annotated Bibliography on Commodity-Intensive versus Subsistence Economies" in Ivan Illich, Shadow Work, pp. 126-127.

9. Quoted in Bodley, Victims of Progress, p.167.

10. See Kelly Jon Morris, "Cinva-Ram," Shelter (Shelter Publications, 1973), p.68. Incidentally, stabilizing mud may lower its thermal benefit. Mud's insulating effect depends almost directly on its thickness. If the strength that cement adds tempts the builder to build thinner (say one brick thick, instead of two), then thermal lag — the hours the structure stays cool after sunrise and warm after sunset — will diminish sharply. See William Lumpkins, "A Distinguished Architect Writes on Adobe," El Palacio, 77:4 (1972), pp. 2-10.

11. See Lloyd Timberlake, "Mud Can Make It," p.6.

12. See Jean-Paul Bourdier, "Genius Before Industry," Dwellings, Settlements, and Tradition (University Press of America, 1989), pp. 56-58.

13. Inaccuracies result in defective shelter. Concrete impurely mixed or improperly cast will set badly and, in the desert's temperature extremes, will be particular liable to crack. See Brent C. Brolin, The Failure of Modern Architecture (Sudio Vista, 1976), p. 105.

14. Hassan Fathy makes a point that is relevant around the world: "Really low-cost housing must not need non-existent resources; mud-brick houses are today made all over Egypt without the help of machines and engineers, and we must resist the temptation to improve on something that is already satisfactory." From Architecture for the Poor, p.134.

15. See Julius Nyere, "The Arusha Declaration Ten Years After," Toward a Way of Life that Is Outwardly Simple, Inwardly Rich (Government Printer, Tanzania, 1977).

16. For a discussion and debunking of this myth in relation to architecture, see David

Watkin, Morality and Architecture Revisited (John Murray, 2001), pp. 9-11.

Case Study: Woodless Construction: Saving Trees in the Sahel

1. The average population growth rate in the region is between 2.5 and 3.5 percent per annum, and as high as 7 percent in the urban areas. See "An Urbanising World," Global Report on Human Settlements (UNCHS, 1996).

2. A Canadian NGO, ISAID, invited DW to demonstrate woodless construction in the context of a small, rural development and environmental management project in Niger in 1980.

Chapter 3: Sustainable Building As Appropriate Technology

Case Study: Casas Que Cantan: Community Building in Mexico

1. See Athena Swentzell Steen and Bill Steen, The Beauty of Straw Bale Homes (Chelsea Green Publishers, 2000).

Chapter 4: Down to Earth Technology Transfer

1. See Zopilote, 1998.

2. See Zopilote, 2000.

3. See Note 1.

4. See Note 2.

Chapter 5: Tell, Show, Do: Teacher Training Programs for Tomorrow's Housing Solutions

1. All information for Alto Mayo was acquired through ITDG's publication Developing Building for Safety Programs (ITDG, 1995); case study notes and external evaluation notes by Dr. Patricia Richmond; furnished by ITDG, Rugby, England.

2. Information was provided by Oxfam, in Oxford, England in the form of project notes by Jolyon Leslie and final evaluation by Andrew Coburn; as well as case study information from Developing Building for Safety Programs (ITDG, 1995).

Chapter 6: Sustainable Settlements: Rethinking Encampments for Refugees and Displaced Populations

1. From 60 percent to 30 percent between 1993 and 1998, poverty being defined as where a family cannot satisfy the bare necessities of life. Vietnam Living Standards

Survey 1997-1998, Hanoi, 1999, Vietnam General Statistical Office.

2. Families whose income levels fluctuate above and below an acceptable minimum level during periods of natural disaster and economic crisis, Vu Tuan Anh, GSO, Vietnam, 1999.

3. Tran Nhon, Vice-Minister, Ministry of Water Resources, Hanoi, 1998.

4. Development Workshop is a not-for-profit organization working to strengthen communities' capacity to meet their shelter needs in a sustainable manner; figures taken from DW VN: database on house, improvement, and strengthening costs.

5. World Bank report: "Viet Nam: Attacking Poverty," November 1999.

6. Commissioned by DW and written, produced, and performed by the Hanoi Puppet Theater Company.

Glossary

Adobe: A sun-dried mud brick (an adobe). Also the style of building with adobes (adobe construction).

Agave stalks: The dried stalk of the succulent plant "agave" (also known as the "century plant").

Agroforestry: A perennial food system that is modeled on forest ecosystems.

Berm: Mounded-up earth, generally used to divert water or to backfill against a wall.

Biomass: Vegetation or other organic material.

Bioreserve: A place preserved or set aside as habitat for plants and animals.

Bond beam: A continuous beam, generally of wood or concrete and steel, placed on top of a wall to give it structural integrity, which is especially important for seismic areas.

Catenary vault: A single curvature roof structure (usually over a square or rectangular room), with the cross-section being a catenary curve. A catenary vault has no bending stresses within the vault itself, making it extremely stable.

Cob: Clay, sand, and straw mixed with water to form a stiff mud, used to create monolithic earthen buildings.

Cold joint: A joint formed when a unit of material (such as concrete or earth) is poured or built next to a previously completed unit of material and is not physically or chemically bonded to it. Unless properly designed for, such joints may cause structural or other problems in a structure.

Cross-bracing: Diagonal bracing, used to provide stiffness to a wall or other building system.

Daylighting: Design strategies where sunlight is the main source of illumination within a building during the daytime.

Dead loads: Loads (weights or forces) on a roof or other structure that do not change over time (the weight of roofing materials, etc.).

Dry-stacking: Stone walls laid without mortar. It is highly skilled labor and demands rigorous attention to secure stone placement.

Ecotourism: A type of tourism designed to support indigenous peoples and to minimize damage to ecosystems.

Ecovillage: An intentional community committed to ecological principles that integrate ecological design with social and economic systems.

Fly ash: The ash that remains in smoke stacks after burning coal.

Formwork: Materials used to temporarily support another material while it is being installed. For example, wooden formwork is commonly used in straw-clay and rammed earth construction.

Grade beam: Similar to a bond beam, a grade beam provides continuous structural support at the base of a wall.

Greywater: Water from showers, washing, laundry, etc. Can be used for watering plants after treatment. Not to be confused with blackwater, which is anything that comes from the toilet.

Gusset plate: Thin, stiff sheet material (like plywood) attached over joints between truss elements (for example) to provide structural stability.

Infill: A type of building system where material is used to fill in between posts. For example, wattle and daub.

Jig: A device used to stabilize components to enable standard cutting or assembly. Especially useful for repetitive tasks.

Joists: Horizontal structural members used to support floors or ceilings.

Knee bracing: Diagonal bracing between the top of a post and a beam that provides lateral stability.

Lateral load: A horizontal load (wind, earthquake, etc.) on a building.

Lintel: A small beam above a door or window to carry the wall load above to the sides of the door or window.

Live load: Loads that change over time (people, snow, etc.)

Load-bearing capacity: The amount of weight a given material can carry before failing. Generally denoted by weight per area (i.e. pounds per square inch).

Micro-enterprise: Small local enterprises, usually with minimal financial investment and serving a geographically restricted area.

Micro-loan: Small loans (often as small as ten dollars) to entrepreneurs to help them get started in business.

Mudbrick: See Adobe

Papercrete: A mixture of re-pulped paper, cement, and sand; used to make building blocks or poured walls.

Permaculture: A term (derived from "permanent" and "agriculture"), coined by Bill Mollison, to describe the design of human landscapes that are agriculturally productive and socially stable, and which are based on principles and patterns derived from nature.

Photovoltaic panels: Silicon-based materials that derive electrical energy from direct solar gain.

Piers: Foundation elements (usually of stone or concrete) that support point loads like posts or beams

Pins: In timber frame construction, wooden dowels used to keep mortise and tenon joints from slipping. In straw bale construction, internal or external elements (usually of wood, steel, or bamboo) used to stiffen the straw bale walls.

Pole foundation: A foundation system, common in tropical areas or steeply sloped areas, where poles are placed in the ground to support beams up in the air.

Pumicecrete: A mixture of volcanic pumice (porous stone) and cement to create an insulating yet water resistive material for walls and foundations.

Purlin: A horizontal roof member that spans rafters or beams and supports the roofing material.

Rammed earth: Slightly moistened pure mineral earth (clay, sand, and aggregate) rammed in between forms. Can be stabilized with cement.

Retrofit: Improving a building's structural performance by changing existing or adding additional elements.

Ring beam: Similar to a grade beam or bond beam, but referring to circular buildings.

Rubble trench foundation: A trench dug below frostline (the level at which the ground freezes in winter), which is then filled with clean stone ballast in which drainage pipe has been imbedded to ensure positive drainage. The compacted stones transfer the loads of the building to the earth below, while allowing moisture to be removed.

Running bond: A style of stone- or bricklaying where the bricks of a subsequent layer are placed to cross the joints of the previous layer.

Scoria: A type of porous volcanic stone.

Shadescreen: Woven material used to create shade from the sun.

Shear forces: A force or series of forces that act perpendicular to the longitudinal axis of a structural member (such as a post) or system (such as a wall).

Shuttered stone masonry: Mortared stone made into walls between movable forms.

Slip form: A temporary form, which is removed after wall-building material is placed within it and moved to the next level. Commonly used in straw-clay construction.

Straw bale flakes: Straw bales are made of discrete, compressed bits of straw called "flakes" that can be separated from each other to stuff into cracks, etc.

Straw-clay: A traditional German building system where loose straw is mixed with a clay slip to produce a sticky mass that can be tamped in between forms. Has also been used to make blocks and ceiling treatments.

Stretcher stones: Stones used in wall building to tie the inside and outside of a wall together.

Struts/chords: The elements of a truss. A chord is the top or bottom member that is connected by the struts.

Thermal break: A space left between heat-conducting materials to arrest heat flow.

Thermosiphon: A system (such as a solar water heating system) where water or some other liquid is heated at a point in the system, and where convection (movement) is created by the differing density of the hot and cold liquid.

Through stones: See "Stretcher stones."

Tie stones: See "Stretcher stones" and "Through stones."

Vernacular architecture: An architectural style that develops from the particular climatic and social conditions of a place.

Vertical loading: Weight that comes onto a building from above.

Wattle and daub: Woven sticks (wattle) and earth or lime plaster (daub); used as an infill material.

Wicking bed system (wastewater treatment): A type of greywater treatment system where a bed of pumice is used to wick water from a discharge area. The pumice provides a place for aerobic bacteria to break down wastes. This system also relies of living plants to take up the resulting nutrients.

Resources

Compiled by David Bainbridge, Joseph F. Kennedy, Susan Klinker, and Melissa Malouf

Publications

Abrams, Charles. *Man's Struggle for Shelter*. MIT Press, 1964.

Agarwal, Anil. *Mud Mud: The Potential of Earth-Based Materials for the Third World*. Earthen Publishing, 1981.

Anderson, Bruce, and Malcolm Wells. *Passive Solar Energy: The Homeowner's Guide to Natural Heating and Cooling,* 2nd ed. Brick House Publishing, 1994.

Awontona, Adenrele, and Michael Briggs. "The Enablement Approach and Settlement Upgrading in South Africa." *Tradition, Location, and Community: Placemaking And Development,* Adenrele Awotona and Necdet Teymur, eds. Avebury Press, 1997.

Bahadori, M.N. "Passive Cooling Systems in Iranian Architecture." *Scientific American* 144-154, 1978.

Baker, Laurie. "Cost-Reduction Manual" and "Architecture and the People." *Architecture and Urbanism,* A+U, No. 363, December 2000. A+U Publishing Company, Japan.

Bhatia, Gautama. Laurie Baker: *Life, Works, and Writing.* Penguin Books, 1991.

Bainbridge, D.A. *The Integral Passive Solar Water Heater Book.* PSI, 1981.

Bourgeois, Jean-Louis, and Carollee Pelos. *Spectacular Vernacular.* Aperture Foundation, 1996.

Burgess, Rod. "Self-Help Housing Advocacy: A Curious Form of Radicalism. A Critique of the Work of John Turner." *Self-Help Housing: A Critique.* P. Ward, ed. Mansell, 1982.

Butti, K., and J. Perlin. *A Golden Thread.* Cheshire Books, 1980.

Chambers, R. *Rural Development: Putting the Last First.* Longman Scientific, 1983.

Chiras, Daniel. *The Natural House: A Complete Guide to Healthy, Energy-Efficient, Environmental Homes.* Chelsea Green, 2000.

Cuny, F. *Scenario for a Housing Improvement Program in Disaster-Prone Areas.* Intertect, 1978.

Davis, Ian. *Shelter after Disaster.* Oxford Polytechnic Press, 1978.

Davis, Shelton H., ed. "Indigenous Views of Land and the Environment." World Bank Discussion Paper. The World Bank, 1993.

Dethier, Jean. *Down to Earth — Adobe Architecture: An Old Idea, a New Future.* Facts on File, 1981.

Dudley, Eric. *The Critical Villager: Beyond Community Participation.* Routledge, 1993.

Drengson, Alan, and Duncan Taylor. *Ecoforestry: The Art and Science of Sustainable Forest Use.* New Society Publishers, 1997.

Dudley, Eric, and Ane Haaland. *Communicating Building for Safety: Guidelines for Methods of communicating technical information to local builders and householders.* Intermediate Technology Publications, 1995.

Eckholm, E. *The Other Energy Crisis.* Worldwatch Institute, 1975.

Elizabeth, Lynne, and Cassandra Adams, eds. *Alternative Construction: Contemporary Natural Building Methods.* John Wiley and Sons, 2000.

Esteva, Gustavo. "Development." *The Development Dictionary: A Guide to Knowledge As Power,* Wolfgang Sachs, ed. Zed Books, 1997.

Evans, Ianto, Linda Smiley, and Michael G. Smith. *The Hand-Sculpted House: A Practical and Philosophical Guide to Building a Cob Cottage.* Chelsea Green, 2002.

Fathy, Hassan. *Architecture for the Poor: An Experiment in Rural Egypt.* University of Chicago Press, 1973.

Fathy, Hassan. *Natural Energy and Vernacular Architecture: Principle and Examples with Reference to Hot Arid Climates.* University of Chicago Press, 1986.

Fass, Simone. "Housing the Ultra-Poor: Theory and Practice in Haiti." JAPA, *Journal of the American Planning Association,* Vol. 53 No. 2. American Planning Association, 1987.

Friedmann, John. *Empowerment: The Politics of Alternative Development.* Blackwell Publishers, 1992.

Gilbert, Alan, and Josef Gugler. *Cities, Poverty, and Development: Urbanization of the Third World.* Oxford University Press, 1992.

Givoni, Baruch. *Man, Climate, and Architecture.* Elsevier, 1969.

Gugler, Josef, ed. *The Urbanization of the Third World.* Oxford University Press, 1988.

Hallsworth, E.G. *Anatomy, Physiology, and Psychology of Soil Erosion.* Wiley and Sons, 1987.

Hamdi, Nabeel. *Housing without Houses: Participation, Flexibility, Enablement.* Van Nostrand Reinhold, 1991.

Holmes, Stafford, and Michael Wingate. *Building with Lime: A Practical Introduction.* Intermediate Technology Publications, 1997.

Houben, Hugo, and Hubert Guillaud. *Earth Construction: A Comprehensive Guide.* Intermediate Technology Publications, 1994.

Isbister, John. *Promises Not Kept: The Betrayal of Social Change in the Third World.* Kumarian Press, 1998.

Jackson, Hildur, and Karen Svensson. *Ecovillage Living.* Green Books, 2002.

Janssen, Jules. *Building with Bamboo.* Intermediate Technology Publications, 1995.

Kahn, L. *Shelter.* Shelter Publications, 1973.

Kellet, Peter, and Mark Napier. "Squatter Architecture? A Critical Examination of Vernacular Theory and Spontaneous Settlement with Reference to South America and South Africa." *TDSR, Traditional Dwellings and Settlement Review,* Vol. VI No. 11. IASTE, International Association for the Study of Traditional Environments, 1995.

Kennedy, Joseph F., Michael G. Smith, and Catherine Wanek, eds. *The Art of Natural Building.* New Society Publishers, 2002.

Kern, Ken. *The Owner-Built Home.* Charles Scribner's Sons, 1972.

Kerr, B., and S. Cole. *Solar Box Cookers International,* manuals and newsletters.

Khalili, Nader. *Ceramic Houses and Earth Architecture: How to Build Your Own.* Cal-Earth, 1986.

King, Bruce. *Buildings of Earth and Straw: Structural Design for Rammed Earth and Straw Bale Houses.* Ecological Design Press, 1996 (distributed by Chelsea Green).

Komatsu, Eiko, Athena Steen, and Bill Steen. *Built by Hand: Vernacular Buildings Around the World.* Gibbs Smith Publisher, 2003.

Komatsu, Yoshio. *Living on Earth.* Fukuinkan-Shoten Publishers, 1999.

Leaf, Michael L. *Urban Housing in Third World Market Economies.* Produced by the Asian Urban Research Network, the Center for Human Settlements, the Canadian International Development Agency (CIDA), and the University of British Columbia, 1993.

Long, Charles. *The Stone Builder's Primer: A Step-by-Step Guide for Owner-Builders.* Firefly Books, 1998.

Low, Nicholas, Brendan Gleeson, Ingemar Elander, and Rolf Lidskog. *Consuming Cities: The Urban Environment in the Global Economy after the Rio Declaration.* Routledge, 2000.

Lyle, John Tillman. *Regenerative Design for Sustainable Development.* John Wiley and Sons, 1994.

Macdonald, Laura. *Supporting Civil Society: The Political Role of Non-Governmental Organizations in Central America.* MacMillan, 1997.

Margolis, R., and N. Salafsky. *Measures of Success: Designing, Managing, and Monitoring Conservation and Development Projects.* Island Press, 1998.

Martinussen, John. *Society, State, and Market: A Guide to Competing Theories of Development.* Fernwood Publishers, 1997.

McDonough, William, and Michael Braungart. "The Birth of the Sustainable Economy: The Next Industrial Revolution." Earthone Productions, Stevenson, Maryland, 2001 [video].

McHenry, Paul Jr. *Adobe: Build It Yourself.* University of Arizona Press, 1985.

McKnight, John, and John P. Kretzmann. *Building Communities from the Inside Out: A Path Toward Finding and Mobilizing a Community's Assets.* ACTA Publications, 1997.

Myhrman, Matts, and S.O. MacDonald. *Build It With Bales: A Step-by-Step Guide to Straw Bale Construction* (Version Two). Out on Bale, 1997.

Mitchell, Maurice. *Culture, Cash, and Housing.* VSO/IT Publication, 1992.

Mollison, Bill. *Permaculture: A Designer's Manual.* Island Press, 1990.

Moser, Caroline O.N., and Linda Peake. *Women, Human Settlements, and Housing.* Tavistock Publications, 1987.

Murison, Hamish S., and John P. Lea, eds. *Housing in the Third World: Perspectives on Policy and Practice.* St. Martin's Press, 1979.

Nelson, Nici, and Susan Wright. *Power and Participatory Development: Theory and Pratice.* Intermediate Technology Publications, 1995.

Norton, John. *Building with Earth.* Intermediate Technology Publications, 1997.

Norton, John. "Indigenous Housing and the Third World." Ekistics, Vol. 41 Jan-June 1976.

Norton, John. "Woodless Construction." Building Issues, Vol. 9 No. 2., 1997.

Norton, John, and Farokh Afshar. "The Developmental Theory." *The Encyclopedia of Vernacular Architecture of the World.* Cambridge University Press, 1.1.8, 1995.

Olgyay, V. *Solar Control and Shading Devices.* Princeton University Press, 1976.

Oliver, Paul. *Dwellings: The House Across the World.* Phaidon, 1987.

O'Meara, Molly. *Reinventing Cities for People and the Planet.* Worldwatch Institute, 1999.

Pacey, A., and A. Cullis. *Rainwater Harvesting.* Intermediate Technology Publications, 1986.

Packer, B. *Appropriate Paper-Based Technology.* IRED, 1995.

Papanek, V. *The Green Imperative: Natural Design for the Real World.* Thames and Hudson, 1995.

Payne, Geoffrey K. *Urban Housing in the Third World.* Leonard Hill, 1977.

Pearson, David. *Earth to Spirit: In Search of Natural Architecture.* Chronicle Books, 1994.

Peattie, Lisa Redfield. *The View from the Barrio.* The University of Michigan Press, 1968.

Pieterse, Jan Nederveen. "My Paradigm or Yours? Alternative Development, Post Development, Reflexive Development." *Development and Change,* Vol. 29. Institute of Social Studies. Blackwell Publishers.

Rapley, John. *Understanding Development: Theory and Practice in the Third World.* Lynne Rienner Publishers, 1996.

Rudofsky, Bernard. *Architecture without Architects: A Short Introduction to Non-Pedigreed Architecture.* The University of New Mexico Press, 1987 [1965].

Rudofsky, Bernard. *The Prodigious Builders.* Harcourt Brace Javanovich, 1977.

Sachs, Wolfgang. "The Need for the Home Perspective." *The Post Development Reader,* Majid Rahnema and Victoria Bawtee, eds. Zed Books, 1997.

Schor, Juliet. *The Overspent American: Upscaling, Downshifting, and the New Consumer.* Basic Books, 1998.

Spence, R., J. Wells, and E. Dudley. *Jobs From Housing.* Intermediate Technology Publications, 1993.

Srinasan, Lyra. *Tools for Communicating Participation: A Manual for Training of Trainers in Participatory Technology.* United Nations, UNDP (Habitat), 1990.

Steel, James. *An Architecture for the People: The Complete Works of Hassan Fathy.* Whitney Library of Design, 1997.

Steel, James. *Sustainable Architecture: Principles, Paradigms, and Case Studies.* McGraw Hill, 1997.

Steen, Athena Swentzell, David Bainbridge, Bill Steen, and David Eisenberg. *The Straw Bale House.* Chelsea Green, 1994.

Steen, Athena, and Bill Steen. "Building Across Cultures." *Designer/Builder: A Journal of the Human Environment,* Vol. VI No. 10, February 2000. Kingsley Hammett, 2000.

Stokke, Hugo, Astri Suhrke, and Arne Tostensen, eds. *Human Rights in Developing Countries,* Yearbook 1997. Kluwer Law International, 1997.

Stulz, Roland, and Kiran Mukerji. *Appropriate Building Materials: A Catalogue of Potential Solutions.* Swiss Centre for Appropriate Technology, IT, GATE, 1988.

Tipple, Graham, and Kenneth Willis. *Housing the Poor in the Developing World: Methods of Analysis, Case Studies, and Policy.* Routledge, 1991.

Tucker, Vincent. "The Myth of Development: A Critique of the Eurocentric Discourse." *Critical Development Theory,* Ronaldo Munck and Denis O'Hearn, eds. Zed Books, 1999.

Turner, John. "Housing Priorities, Settlement Patterns, and Urban Development in Modernizing Countries." *AIP Journal,* November 1968.

Turner, John F.C. *Housing By People: Towards Autonomy in Building Environments.* Pantheon Books, 1976.

Turner, John, and Rolf Goetze. "Environmental Security and Housing Input." *Ekistics,* Vol. 23

United Nations. *Manual on Self-Help Housing.* New York, 1964.

UNCHS. *Report of the Ad Hoc Expert Group Meeting on the Development of the Indigenous Construction Sector.* United Nations Center for Human Settlements, 1982.

UNCHS. *Global Report on Human Settlements.* Oxford University Press (for the United Nations Center for Human Settlements), 1987.

UNCHS. *Roles, Responsibilities, and Capabilities for the Management of Human Settlements: Recent Trends and Future Prospects.* United Nations Center for Human Settlements, 1990.

UNCHS. *Shelter for All: Global Strategy for Shelter to the Year 2000.* United Nations Center for Human Settlements, 1990.

van Lengen, Johan. *Manual del Arquitecto Descalzo (Manual of the Barefoot Architect).* Editorial Concepto, 1980.

Vélez, Simón, et. al. *Grow Your Own House.* Vitra Design Museum, 2000.

Wojciechowska, Paulina. *Building With Earth: A Guide to Flexible-Form Earthbag Construction.* Chelsea Green, 2001.

World Bank. *Urban Policy and Economic Development: Agenda for the 1990s.* A World Bank Policy Paper. The World Bank, 1991.

World Commission on Environment and Development. *Our Common Future.* Oxford University Press, 1987.

Zopilote Association. *Sustainable Develoment and Natural Building Course Language Notebook.* s.p. Zopilote, 2000.

Organizations

Adobe Alliance

PO Box 1915, Presidio, Texas 79845
Web: www.adobealliance.org
E-mail: simone@adobealliance.org
Telephone: (432) 229 4425
Facsimile: (432) 229 4425.
The purpose of the Adobe Alliance is to build low-cost, energy-efficient housing that is climatically and environmentally compatible; and to fill widespread needs for sustainable, salubrious housing while enhancing the unique landscape of the Big Bend region of West Texas and other desert environments.

Auroville Building Centre

Earth Unit Auroshilpam, 605 101 Auroville, Tamil Nadu, India
Web: www.auroville.org
E-mail: csr@auroville.org.in
Telephone: 91-4136-2168/2277   Facsimile: 91-4136-2057
Research, publications, and information on earth building for development, especially with compressed earth blocks.

Builders Without Borders

119 Main Street, Kingston, NM 88042, USA
Web: www.builderswithoutborders.org
E-mail: mail@builderswithoutborders.org
Telephone: 505-895-5400
Courses, technical consulting, and hands-on projects to serve the underhoused. Focuses on ecological building in cross-cultural contexts.

Building Advisory Service and Information Network (BASIN)

Web: www.gtz.de/basin/.
BASIN provides information and advice on appropriate building technology, and creates links with know-how resources in the world for all those in need of relevant information: government officers, financiers, builders and developers, architects, planners, and producers of building materials who need up-to-date information and advice on the manufacture, performance, and availability of appropriate outputs and technology from around the world, and on the effective management of local resources.

California Earth Art and Architecture Institute (Cal-Earth)

10376 Shangri-La Avenue, Hesperia, CA 92345, USA
Web: www.calearth.org
Telephone: 760-244-0614
Facsimile: 760-244-2201
A research and teaching center focusing on earthbag construction, adobe, and brick domes and vaults.

The Canelo Project

HC1 Box 324, Elgin, AZ 85611, USA
Web: www.caneloproject.org
E-mail: absteen@dakotacom.net

Telephone: 520-455-5548.

The Canelo Project is a nonprofit educational organization that offers workshops on straw bale construction, earthen floors, earth and lime plasters, work-study tours to Mexico, and more.

Center for Alternative Technology (CAT)

Canolfany Dechnoleg Amgen, Machynlleth, Powys SY20 9AZ UK

Web: www.cat.org.uk

E-mail: info@cat.org.uk

Telephone: 01654 702400.

Publications and a demonstration center on a wide variety of sustainability issues.

The Centro Experimental de la Vivienda Económica (CEVE)

Igualdad 3585 - B° Villa Siburu, 5003 Córdoba, Argentina

Web: http://ceve.org.ar

E-mail: basin@ceve.org.ar

Telephone: +54-351-489-4442;1

Facsimile: +54-351-489-4442;1

CEVE is a non-governmental, professional association and member of the Asociación de Vivienda Económica (AVE) in Argentina. Its objectives are the promotion of low-cost housing and the creation of new job sources through technology research and development, training and technology transfer.

CRATerre

Maison Levrat, Rue de Lac B.P. 53, F-38092 Villefontaine Cedex, France

Web: www.craterre.archi.fr

E-mail: craterre@club-internet.fr

Telephone: 33-474-954391

Facsimile: 33-474-956421

Professional school of earth architecture, engineering, and construction; especially for developing countries.

Development Alternatives

B-32 Tara Crescent, Qutab Institutional Area, New Delhi - 110 016, India

Web: www.devalt.org

E-mail: tara@sdalt.ernet.in

Telephone: +91-11-66-5370; 696-7938; 685-1158

Facsimile: +91-11-686-6031

Development Alternatives is a not-for-profit sustainable development enterprise that designs and promotes programs and products which, through the use of alternative technology, contribute to the enrichment of human life. Its symbol, a five-pointed star, depicts a stylized human figure with arms stretched outward, its head pointing skyward, and its feet firmly rooted to the earth.

Development Center for Appropriate Technology (DCAT)

PO Box 27513, Tucson, AZ 85726-7513, USA
Web: www.dcat.net
E-mail: info@dcat.net
Telephone: 520-624-6628
Facsimile: 520-798-3701

The Development Center for Appropriate Technology works to enhance the health of the planet and our communities by promoting a shift to sustainable construction and development through leadership, strategic relationships, and education.

Development Workshop France (DWF)

B.P. 13, 82110 Lauzerte, France
Web: www.dwf.org
E-mail: dwf@dwf.org
Telephone: +33 (5) 63 95 82 34

DWF is a capacity building and action–research organization involved with the implementation of programs in the field. DWF works to develop local capacities to improve living conditions in developing and threatened communities. DWF activities focus on vulnerability reduction and the resolution of human settlement difficulties that result from environmental and demographic change and from man-made and natural disasters.

EcoSur (EcoSouth)

Apdo 107, Jinotepe, Nicaragua
Web: www.ecosur.org
E-mail: ecosur@ibw.com.ni
Telephone: +505-4223325
Facsimle: +505-4223325

Ecosouth, Schatzgutstrasse 9, 8750 Glarus, Switzerland
E-mail: sofonias@compuserve.com

Telephone: +41 556401081

Facsimile: +42-556401081

EcoSur (EcoSouth), the Network for an Ecologically and Economically Sustainable Habitat, promotes a larger scope of activities in the field of ecologically and economically sustainable construction. The network seeks to take advantage of current synergies among countries, persons, and entities active in these fields. EcoSouth focuses its activities in southern countries and propagates South-South technology transfer as well as actual trade where equipment and material is concerned.

The German Appropriate Technology Exchange (GATE)

Dag-Hammarskjöld-Weg 1-5 D-65760 Eschborn, Germany; PO Box 5180, D-65726 Eschborn, Germany

Web: www.gtz.de/gate/

E-mail: gate-basin@gtz.de

Telephone: +49-6196-79-4212

Facsimile: +49-6196-797352

GATE, a part of the Deutsche Gesellschaft für Technische Zusammenarbeit (GTZ), specializes in environmental resource protection and dissemination of appropriate technologies for developing countries.

Global Ecovillage Network

Web: http://gen.ecovillage.org/

The Global Ecovillage Network is a global confederation of people and communities that meet and share their ideas; exchange technologies; develop cultural and educational exchanges, directories, and newsletters; and are dedicated to restoring the land and living "sustainable plus" lives by putting more back into the environment than we take out.

Intermediate Technology Development Group

The Schumacher Centre for Technology & Development, Bourton Hall, Bourton-on-Dunsmore, Rugby CV23 9QZ, UK

Web: www.itdg.org

E-mail: itdg@itdg.org.uk

Telephone: +44 (0)1926 634400

Facsimile: +44 (0)1926 634401

IDTG was founded in 1966 by the radical economist Dr. E.F. Schumacher to prove that his philosophy of "Small is Beautiful" could bring real and sustainable improvements to people's lives. The organization offers publications, consulting, and other services to reduce the vulnerability of poor people affected by natural disasters, conflict, and environmental degradation; and to enable them to make a better living; respond to the challenges of new technologies; and gain access to basic services, such as safe, clean water, food, housing, and electricity.

Pagtambayayong

102 P. del Rosario Ext., Cebu City, 6000 Philippines
E-mail: pagtamba@cnms.net
Telephone: +63-32-2537974
Facsimile: +63-32-4182168

Pagtambayayong — A Foundation for Mutual Aid, Inc. is a non-governmental organization that promotes low-income housing. Its Centre for Appropriate Technology conducts research, production, and training on appropriate building technologies. Pagtambayayong is a member of the Sustainable Building Technologies — Philippines, the network of SBT practitioners.

Shelter Forum

PO Box 39493, Nairobi, Kenya
E-mail: ericm@shelterforum.or.ke
Telephone: +254-(-020)-3753181 / 2
Facsimile: +254 (-020) –3753180

Shelter Forum is a coalition of non-governmental organizations that deal with issues of low-cost housing shelter in Kenya. Its main goal is to enhance access to affordable shelter for all, particularly the poor, among whom the most vulnerable are women and children, through advocacy, extension, and networking.

SKAT

Vadianstr. 42, CH-9000 St. Gallen, Switzerland
Web: http://www.skat.ch
E-mail: info@skat.ch
Telephone: +41-71-2285454
Facsimile: +41-71-2285455

The Swiss Centre for Development Cooperation in Technology Management (SKAT) is a documentation center and consultancy group active in the field of development cooperation. It provides technical back-up to balanced development which is institutionally, ecologically, and economically sound as well as technically realistic.

Contacts

Liora Adler

Web: www.larcaravana.org

E-mail: liora@lacaravana.org

David A. Bainbridge

Associate Professor, Sustainable Management

United States International College of Business

Alliant International University, Rm. B-1

10455 Pomerado Road, San Diego CA 92131, USA

Web: www.alliant.edu/faculty/bainbridge.htm

www.sustainableenergy.org/resources/technologies/solar_passive.htm

www.ecocomposite.org

Telephone: 858-636-4616

Facsimile: 858-635-4528

Jean-Louis Bourgeois dit Baber Maiga

PO Box 526, El Prado NM 87529, USA

E-mail: jeanlouisbourgeois@yahoo.com

Telephone: Taos, New Mexico 505-751-1282

New York: 212-242-4984

Djenne, Mali: 223-420-066

Cameron M. Burns

Rocky Mountain Institute

1739 Snowmass Creek Road, Snowmass CO 81654-9199, USA

Web: http://rmi.org

E-mail: cameron@rmi.org

Telehone: (970) 927-3851 (general)

(970) 927-7338 (direct)

Facsimile: (970) 927-3420

Paul Cohen

Memento Cottage

6 Dreyer Avenue, Kommetjie 7975, Cape Town, South Africa

E-mail: vildev@iafrica.com

Telephone/Facsimile: +27-21-783-4007

Brendan Conley

1230 Market Street #131, San Francisco CA 94102, USA

E-mail: brendanconley@lycos.com

Jean D'Aragon and Rosa Fernandez

2250 St. Jacques Ouest, Montreal, Quebec, Canada H3J 2M7

E-mail: jdarag@po-box.mcgill.ca

Telehone: 514-935-9293

Darrel DeBoer

1835 Pacific Avenue, Alameda CA 94501, USA

Web: www.deboerarchitects.com

E-mail: darrel@deboerarchitects.com

Phone: 510-865-3669

Flora Gathorne-Hardy

E-mail: pwojciechowska@hotmail.com

Dr. Owen Geiger

PO Box 354, Radium Springs NM 88054 USA

E-mail: bwb@zianet.com

mail@builderswithoutborders.org

strawhouses@yahoo.com

Phone: 505-526-2559

Fax: 505-895-5400

Joseph F. Kennedy

99 Sixth Street, Santa Rosa CA 95401, USA

Web: www.ecodwelling.org

www.builderswithoutborders.org

E-mail: jkennedy@newcollege.edu

Telephone: (707) 568-3092

Susan Klinker

903 3rd Avenue, Salt Lake City UT 84103, USA

E-mail: susan-klinker@hotmail.com

Phone: (801) 355-9208

Judy Knox

Out on Bale (un)Ltd.

2509 N. Campbell Avenue #292, Tucson AZ 85719, USA

E-mail: judyknox42@aol.com

Phone: (520) 622-6896

Paul Lacinski

Greenspace Collaborative

PO Box 107, Ashfield MA 01330 USA

E-mail: paul@greenspacecollaborative.com

Phone: (413) 628-3800

Kelly Lerner

One World Design

744 W. 12th Avenue, Spokane WA 99204, USA

Web: http://www.one-world-design.com

E-mail: klerner@one-world-design.com

Phone: (509) 838-8812

Ayyub Malik

1 Julius Court, Justin Close, Brentford, Middlesex, London TW8 8QH UK

Web: www.others.com/ayyub.htm

E-mail: malik@yyub.fsnet.co.uk

Phone/Fax: +44 20 8568 1237

Melissa Malouf

PO Box 2858, Crested Butte CO 81224, USA

E-mail: melmalouf@yahoo.com

Phone: (970) 349-0161

Clare Cooper Marcus

Professor Emerita, University of California

Principal, Healing Landscapes

2721 Stuart Street, Berkeley CA 94705, USA

E-mail: clare@mygarden.com

Phone/Fax: (510) 548-2904

Professor Dr. Ing. Gernot Minke

Building Research Institute (BRI)

University of Kassel, Menzelstra?e 13, 34109, Kassel, Germany
E-mail: feb@architektur.uni-kassel.de
Phone: +49 561-804-5315, -5312
Fax: +49-561-804-5428

John Norton

Development Workshop
BP 13, 82110 Lauzerte, France
Web: www.dwf.org
E-mail: dwf@dwf.org
Phone: +33-5-63-95-82-34

Michelle Nijhuis

38552 Pitkin Road, Paonia CO 81428, USA
E-mail: nijhuism@yahoo.com
Phone: (970) 527-3793 (office)

Kathryn Rhyner-Pozak

Grupo Sofonias
Schatzgutstr. 9, 8750 Glarus, Switzerland
E-mail: sofonias@compuserve.com
kpozak@bluewin.ch

David Riley

Department of Architectural Engineering
104 Engineering Unit A, University Park PA 16801, USA
Web: www.engr.psu.edu/greenbuild
E-mail: driley@engr.psu.edu
Phone: (814) 863-2079
Fax: (814) 863-4789

Alex Salazar

PO Box 71072, Oakland CA 94612, USA
E-mail: alexsalazar1138@hotmail.com

Bill Steen

The Canelo Project
HC1 Box 324, Elgin AZ 85611, USA

Web: www.caneloproject.org
E-mail: absteen@dakotacom.net
Phone: (520) 455-5548

Simone Swan

Adobe Alliance
PO Box 1915, Presidio TX 79845, USA
Web: http://adobealliance.org
E-mail: simone@adobealliance.org
Phone/Fax: (432) 229-4425
Voicemail: 1-800-359-6677X77

Alfred von Bachmayr

Von Bachmayr Architects
1406c Bishops Lodge Road, Santa Fe NM 87506, USA
E-mail: vbarch@earthlink.net
Phone: (505) 989-7000
Fax: (505) 984-1479

Ivan Yaholnitsky

Managing Director
Bethel BCDC
PO Box 53, Mt. Moorosi 750, Lesotho, Africa
Web: www.leo.co.ls/bbcdc
E-mail: bbcdc@uuplus.com
Phone (266) 5-874-2991

Index

JOSEPH F. KENNEDY

J OSEPH F. KENNEDY is a designer, builder, writer, artist and educator. Working with pioneers in the field, he has been at the forefront of ecological design and construction for the past seventeen years. He is cofounder of Builders without Borders and teaches in the EcoDwelling program at New College of California. Joseph is a co-editor of *The Art of Natural Building* (New Society 2002) and teaches and works globally.

If you have enjoyed *Building Without Borders,* you might also enjoy other

BOOKS TO BUILD A NEW SOCIETY

Our books provide positive solutions for people who want to make a difference. We specialize in:

**Sustainable Living • Ecological Design and Planning • Natural Building & Appropriate Technology
New Forestry • Environment and Justice • Conscientious Commerce • Progressive Leadership
Educational and Parenting Resources • Resistance and Community • Nonviolence**

For a full list of NSP's titles, please call 1-800-567-6772 or check out our web site at:
www.newsociety.com

New Society Publishers

ENVIRONMENTAL BENEFITS STATEMENT

New Society Publishers has chosen to produce this book on Rolland Enviro 100, recycled paper made with 100% post consumer waste, processed chlorine free, and old growth free.

For every 5,000 books printed, New Society saves the following resources:[1]

53	Trees
4,837	Pounds of Solid Waste
5,322	Gallons of Water
6,942	Kilowatt Hours of Electricity
8,793	Pounds of Greenhouse Gases
38	Pounds of HAPs, VOCs, and AOX Combined
13	Cubic Yards of Landfill Space

[1] Environmental benefits are calculated based on research done by the Environmental Defense Fund and other members of the Paper Task Force who study the environmental impacts of the paper industry.

NEW SOCIETY PUBLISHERS